27 50

CLIO WIRED

Clio Wired

THE FUTURE OF THE PAST
IN THE DIGITAL AGE

Roy Rosenzweig

Introduction by Anthony Grafton

COLUMBIA UNIVERSITY PRESS   NEW YORK

COLUMBIA UNIVERSITY PRESS

Publishers Since 1893

NEW YORK   CHICHESTER, WEST SUSSEX

Library of Congress Cataloging-in-Publication Data

Rosenzweig, Roy.

Clio wired : the future of the past in the digital age / Roy Rosenzweig ; introduction by Anthony Grafton.

p. cm.

Includes bibliographical references and index.

ISBN 978-0-231-15086-6 (cloth : alk. paper)—ISBN 978-0-231-15085-9 (paper : alk. paper)—ISBN 978-0-231-52171-0 (electronic)

1. History—Computer network resources.   2. Historiography—Philosophy. 3. Historiography—Methodology.   4. Historiography—Forecasting.   5. Digital media. 6. Internet.   7. World Wide Web.   8. History—Study and teaching (Higher). 9. History—Research.   10. Historians.   I. Title.

D16.117.R67   2011

902.85—dc22   2010012603

To Mae Rosenzweig

Rethinking History in New Media

Practicing History in New Media: Teaching,
Researching, Presenting, Collecting

Surveying History in New Media

Roy Rosenzweig:
Scholarship as Community

ANTHONY GRAFTON

R OY ROSENZWEIG DIED, MUCH TOO YOUNG, IN 2007. Those who knew and loved him—an astonishingly varied group of colleagues and students, neighbors and e-mail correspondents—will feel the loss for years to come. Yet we can still hear his voice. Like many other fine historians, he continues to teach through his books, which convey his passion for reconstructing with learning, craft, and style the experiences of ordinary men and women. He continues to make himself heard in other ways as well, especially through the vast range of collective enterprises, from Web pages to institutions, to which he gave so much of his energy and intelligence. Every time a high school or college student or a teacher clicks on the *History Matters* or *World History Matters* Web sites; every time an independent scholar reads an article or review in the newest issue of the *American Historical Review* without having to pay; every time someone curious about America and the world browses the September 11, 2001 Archive now held by the Library of Congress, he or she hears Roy's voice and profits from his passions: his love of history, his belief that everyone should share it, and his uncanny ability to make the dream of universal access come partly true. Roy never wrote a best-seller or a conventional textbook. But he devised new methods for both public history and historical education, and by doing so he reached a vast audience outside as well as inside the profession. The collaborative projects that he led, moreover, continue to grow and change, as those in charge of them—many of whom he helped to train—work out technical improvements and take in new materials and points of view.

Academic success came quickly for Roy. He graduated from Columbia in 1971, magna cum laude, and won one of the university's prized Kellett Fellowships, which took him to St. John's College, Cambridge. Moving from the British to the American Cambridge, Roy published his first articles—characteristically, in *Labor History* and *Radical America*—long before he finished his thesis. He received his doctorate in 1978, and even in those years of academic famine, his career went well. In 1981 he began teaching at George Mason University, where he would spend the rest of his academic life. And in 1983 Cambridge University Press published a revised version of Roy's dissertation: *Eight Hours for What We Will: Workers and Leisure in an Industrial City, 1870–1920*. Still in print almost three decades after its first appearance, the book traces the ways the working men and women of Worcester, Massachusetts gained free time and what they made of it. Inspired in part by E. P. Thompson, *Eight Hours for What We Will* remains a powerful and provocative study of how ordinary men and women shape their world. It seems characteristic of Roy that at a time when quantitative history rode high and the analysis of census records seemed to hold illimitable dominion over all, he reconstructed the daily life of the working-class saloon. He argued, with characteristic force and clarity, that the culture that workers created in their new free time was anything but a trivial subject, even if it was not—as romantic left historians might have wished—centered on absolute opposition to existing power structures. Roy's imaginative exploration of a vast range of records led him into many new realms. He teased out the ways different immigrant groups settled, worked, and built communities—a form of analysis that would find multiple echoes a generation later in the new immigration history of the 1990s and after.

Roy, in short, was a skillful and productive scholar who touched all the professional bases. But his life and work as a historian were anything but conventional. A New Yorker who grew up in Bayside, Queens, playing stickball and relying on the New York Public Library, he was schooled in the great tradition of American Marxism—the Groucho tendency—by his father (as a student at Columbia, he named his hyperactive cat Rufus T. Firefly). His life-long commitment to the political Left was accompanied by an equally strong allergy to all forms of pomposity and pretension. Both made him incapable of casting his lot with those who held power in his discipline and his profession. Though Roy studied at two of America's best departments of history, he found relatively little intellectual nourishment in their formal courses and seminars. Long after he left Harvard, one of the *Radical History Review* columns that he wrote, with Jean-Christophe Agnew, under the pen name R. J. Lambrose, included a revealing little story: "Some years ago, a friend returned from a job interview with Harvard's American historians. When asked whether he had 'knocked them dead,' he replied: 'No such luck. They were dead when I got there.'" The words may not have been Roy's, but the attitude they express surely was. He read and explored the world of scholarship and found rich resources there, from E. P. Thompson's vision of how to write the history of the working class, which he encountered at Cambridge, to the computer, which he may have used for the first time when he took Stephan Thernstrom's graduate course. But he selected and applied these tools, and others, in ways profoundly his own.

What shaped Roy, more than any seminar, were his experiences in the urban world outside the university—from the summer he spent in a shoe factory in Brooklyn, which taught him what industrial work really meant to those who performed it, to the early morning sessions he put in on Whitehall Street, outside New York's Selective Service Center, trying to help young men who lacked his advantages avoid the draft and Viet Nam. He never lost the conviction he gained there: that intellectuals, and historians in particular, needed to be continually aware of the political, social, and economic circumstances in which they and others pursued their craft.

As Roy worked to come to terms with what he had seen, he also learned, intensively, from friends and colleagues. He was still in graduate school when he took up what became a lifelong practice of collaboration—a radical innovation at the time. In those days historical work, whatever its methodology, was usually monastic in its form. Each scholar worked for him- or herself, locked away in a carrel, engaged in a desperate struggle to master the sources before being overcome by melancholy or crippled by writer's block. This model of intellectual work never made any sense to Roy, and with Agnew and other friends at Harvard and elsewhere, he soon found ways to give rigorous scholarship a rich and varied social dimension. As the Columbia historian Elizabeth Blackmar, an old friend of Roy's, recently recalled, "What do you do when you don't know what you are doing? You organize a reading group; you form a collective to produce a journal, you make sure that all of your friends know each other—whether in person or as legends. You give other people drafts of your work to read and read theirs and talk to them. Roy helped us all collectively to gain the confidence to do our creative work, and he helped many of us find jobs, housing, roommates, and life-long friends." Projects took shape, one after another: a collaborative book on experimental methods in history teaching, based on more than eighty individuals' experiences; documentary films on Mission Hill, a working-class neighborhood in Boston, and on the social and ecological contexts of public waterworks in the United States.

Above all, there was *Radical History Review.* Like the talented young people in a movie of the thirties, Roy and his friends basically found an old barn, improvised a set and props, and put on a show—one that continues to this day. Though many helped to bring this extraordinary journal into being, Roy played a crucial role, and he helped to edit *RHR*—and to rescue it when organizational or financial disaster loomed—for decades. Like its slightly younger British cousin *History Workshop*, *Radical History Review* did several things at once. It sponsored talks, workshops, and conferences: open intellectual forums that magnetically attracted the gifted young. It offered a partial remedy for the seeming sterility of established forms of historical scholarship. Often, it provided an intellectual home for inquiries, methods, and issues that could not, in those days, be pursued in the *American Historical Review* and other established periodicals; sometimes, it became the forum for debates when such central themes as gender were not pursued systematically enough. It kept up a running, sharply critical commentary on the larger political and economic scene, especially on the ways market considerations and market language were transforming the university, and on the splendors and miseries—mostly the miseries—of the historical profession. When right-wing revisionists sang the praises of Nixon or when left-wing historians came under attack, *Radical History Review* was there to state the other side.

At its best, *RHR* rebutted the flatulent rhetoric of reaction with a withering sarcasm far more effective than any formal argument could have been:

Last October, on the twentieth anniversary of the March on the Pentagon, a group of ex-New Lefties held a "Second Thoughts Conference" in Washington, D.C. There, David Horowitz, Peter Collier, Robert Leiken, Ron Radosh and others ritually recanted "the self-aggrandizing romance with corrupt Third Worldism," the "casual indulgence of Soviet totalitarianism," and the "hypocritical and self-dramatizing anti-Americanism" in which they had participated during their youth. And there, Hilton Kramer, Norman Podhoretz, and Irving Kristol just as ritually refused them absolution. We had planned a comment on this affair but then thought better of it.

Radical History Review shaped scholarship and teaching for a generation and more.

It also gave Roy a rapid training, by immersion, in the practical work of academic organization and publication. He applied the editorial skills he learned with tireless generosity for the rest of his life, not only on the *Review* but also as a coeditor of Temple University Press's series "Critical Perspectives on the Past," which eventually included almost three dozen monographs and collections of essays; as coeditor (with Jean-Christophe Agnew) of *A Companion to Post-1945 America* (Oxford: Blackwell, 2002); and in other venues where he showed his passion for making the back of every triptych as perfect as its front. For *Who Built America?*, a passionate and innovative textbook of social history, Roy served as an editor or author on all three editions. Only his colleagues knew how extensive his input was, even when he initially held the mysterious title of "consulting editor." As Stephen Brier explained in the preface to the first edition, Roy read and commented critically on every chapter in this enormous work, calling for improvements not only in individual parts but also in the presentation of the whole. He convinced the other editors to include a much wider range of first-person documents to enhance the book's flavor and texture, and he continued to help and advise as they did so.

The same critical skills, and the same passion for persuading others to surpass themselves, became central to Roy's practice as a teacher—part of his calling that mattered deeply to him, and to which he devoted himself with a straightforward intensity that is, to say the least, not always shown by senior historians who rival his proficiency and productivity as a scholar. Roy saw critical reading as a craft, one that he practiced with as much flexibility, engagement, and passion when he read freshman papers as when he reviewed colleagues' books. It became a vital part of what he passed on to his students. As Elena Razlogova recalled,

> Roy taught me how to comment on other people's work. When he showed me how to rewrite completely one of my less successful drafts, in a 5-page single-spaced line-by-line commentary, he was quite direct and at times sar-

castic. To one of my wilder propositions he responded, "I am prepared to believe that this is the case, but the claims here seem to rest on two anecdotes." Yet he was also kind—he also used, quite without foundation, words "perceptive," "well-written," and "wonderful," the latter three times. I'm not sure Roy was capable of writing comments that were not detailed—he gave such thorough responses not just to dissertation and book chapters but also to papers he assigned in . . . class.

Roy's love of working with others carried over into his most substantial scholarly projects. Many historians work in teams on textbooks, exhibitions, and other projects aimed at a large public. But relatively few write research-based articles and books together. Roy found it natural to work with others when he spelunked in archives and crafted the facts and stories he mined there into forceful, elegant prose. For much of his career, in fact, he was a serial collaborator. Through the 1980s, he researched and wrote *The Park and the People,* a comprehensive and innovative history of Central Park, with Elizabeth Blackmar. The loneliest of all the historian's activities became, for Roy, another form of sociability: "Thanks to Roy's genius for collaboration," Blackmar writes, "we were able to laugh and argue our way through to producing a book"—a model of scholarly creativity as attractive as it is unusual. *The Park and the People* included the usual suspects, such as Robert Bowne Minturn and August Belmont, Frederick Law Olmsted and Robert Moses. But it also brought the New Yorkers whom Central Park displaced and the workers who built it on stage alongside the patricians, architects, and planners. Designers and politicians, patricians and entrepreneurs, the eight-hour day and public transport, baseball-playing boys and "wheel-women" on bicycles all played their parts in this complex, multilevel analysis, just as all of them had helped to make and remake Central Park's natural and cultural spaces. *The Park and the People* won five prizes for its wealth of new material, new perspectives, and vivid prose. Roy carried out another prize-winning investigation of new fields, *The Presence of the Past: Popular Uses of History in American Life*, in partnership with David Thelen. And he wrote his last book, *Digital History: A Guide to Gathering, Preserving, and Presenting the Past on the Web*, with Daniel Cohen. This exceptionally lucid, informative, and witty book is much more than a guide. It traces the history of the History Web from "walking city" to "sprawling megalopolis"; pays tribute to the pioneers—amateurs as well as professionals; and offers clear instructions, well-chosen screen shots, and wise reflections to help novices setting out to home-stead new parts of the territory.

Three themes intersected, often in unexpected ways, in Roy's later life: computers, institution building, and—above all—community. He was one of the first historians to not only join the PC revolution but also master new tools for programming and analysis of data, as these became available, gen-eration after generation. His first article in the field, "Automating Your Oral History Program: A Guide to Data Base Management on a Microcomputer," appeared in the *International Journal of Oral History* as early as November 1984. He stayed engaged, over the decades, with one technology after an-other, from the CD-ROMs of the 1990s to the World Wide Web, where he soon made himself at home. Roy rapidly established himself as an extraordi-narily creative user of the new technologies. His skills at assembling and ana-lyzing data underpinned his later books. The databases he created to link re-cords of different kinds enabled him and Blackmar to tease apart the layers of Central Park's history and tell the forgotten "stories of the African-American, Irish and German households that had resided on the land."

At the same time, Roy devised, raised funds for, and directed projects that showed, for the first time, what the computer could do for public history—above all, those extraordinary Web-based resources for history teaching, *Lib-erty, Equality, Fraternity* and *History Matters,* which offer users not only a vast range of primary sources in many media but also imaginatively conceived and deftly executed suggestions about the most productive ways of making them speak. In 2003, the Rockefeller Foundation and the National Humani-ties Center awarded him the second Richard W. Lyman Award for "outstand-ing achievement in the use of information technology to advance scholarship and teaching in the humanities"—clear public recognition of his pioneering role and of the importance of his example.

Institutions mattered to Roy from the start of his career. Though he at-tended and gave papers at professional meetings, he criticized the traditional organizations as stiff, hierarchical, and ungenerous. Their senior members looked down, apparently with equal lack of sympathy, on the women, people of color, and gay scholars who were entering the historical profession. When the distinctive academic prosperity of the 1960s, which had chiefly benefited white males, came to an end, the great men responded to the economic and personal misery that faced young historians by calling for cuts in graduate enrollments. Only those who turned their backs on the traditional structures, it seemed, could build institutions that might promote a new, more democratic intellectual life. Hence *Radical History Review* and so much else.

Roy never lost his passion for democracy. But over time, as he and other members of his generation became recognized as master historians in their turn, they found places—and began to see a new potential—in the older, larger institutions. In the years around 1970, Roy was one of many driven to despair by the old guard of the American Historical Association. By 2000, Eric Foner, an older friend and intellectual ally, was president of the AHA. Roy became very engaged in the work of the association, which he served, for three years, as a supremely effective Vice President for Research. Urgent and eloquent as ever, he convinced his colleagues that the AHA's sometimes-stodgy ways needed to change. He persuaded the association's council to open universal Web access to its journal, the *American Historical Review*, and its other publications, and worked with colleagues to modernize the association's annual meetings.

But it was at George Mason, the raw but aspiring university where Roy taught for more than a quarter of a century, that he did his most distinctive construction job of all. Starting in the early 1990s, Roy created the Center for History and New Media. At the beginning, the grand name referred only to his office in the history department. Sustained by vision, commitment, and the rhetorical abilities that made him a consistently successful applicant for grant support, Roy designed projects that showed the potential of digital history—at least when designed and brought into being by a master researcher and storyteller. More remarkable, he devised a structure to support these efforts, one that proved capable of seemingly endless growth and modification. As the center took on equipment and workers, it spread into vacant offices, then moved into a leaky trailer, and finally occupied a large suite of rooms in George Mason's Research 1 building. This glistening, purpose-built research facility houses, in addition to the center's staff of more than forty, the Center for Earth Observing and Space Research and the Computational Materials Science Center. It seems a strange place to find a nest of historians, but, in fact, it makes a natural home for the house that Roy built.

As large as George Mason's history department, the center is now directed and run by a team of historians who resemble Roy in their mastery of traditional narrative as well as contemporary digital technologies—along with some full-time specialists in programming and graphic design. Its projects include well-established educational Web sites like *History Matters* and the newer, award-winning *Historical Thinking Matters*; digital collections and exhibitions of original materials such as the *Bracero Archive* and *The Gulag: Many Days, Many Lives*; and new research tools like Zotero, which scholars and graduate students around the country use to collect and manage their primary sources. Roy noted long ago that as digital sources became available in vast quantities, historians would have to learn new ways both to find the relevant materials effectively concealed by the sheer mass of data that confronted them and then to identify meaningful patterns in the evidence. The center has embarked on a study of available methods for text mining and the ways some historians now apply them, and that many more will master in the future.

Dedicated to collaborative enterprises, the center—unlike most traditional history departments—is housed in open-plan offices designed to facilitate discussion. It attracts and benefits from the formidable energies and creativity of George Mason's graduate and undergraduate students, as well as members of the history department and its own full-time staff. In a moment of retrenchment in the humanities, the center—along with allied projects like the American Social History Project and the Center for Media and Learning at the City University of New York—marks what may be the first full-scale new model of historical research and teaching to take shape since the introduction of the seminar system in the late nineteenth century and the creation of the National Archives not long after. With characteristic generosity, Roy always expressed his respect for those unlikely role models, the bearded gentleman scholars who made history a profession in the decades around 1900. They had accepted responsibility for education as well as for research, and for the collection and publication, as well as the interpretation, of historical sources. In creating the center, he realized their vision in a new way—and unexpectedly showed the young historians of the twenty-first century that they could and should emulate the best qualities of their professional grandfathers.

The last vital thread in Roy's work—and the one that pulled the other two together at many points—was his conviction that history embraces not only what historians at universities think and write but also all the ways people see, recall, and understand their pasts. From early in the 1980s, at a time when scholars were only beginning to develop an interest in museums and other public collections, Roy was examining the presentation of the past in magazines and museums and encouraging colleagues to do the same. The collective inquiries that he mounted with colleagues yielded two collections of essays, *Presenting the Past: Essays on History and the Public*, coedited with

Susan Porter Benson and Stephen Brier, and *History Museums in the United States: A Critical Assessment*, coedited with Warren Leon. Above all, in *The Presence of the Past: Popular Uses of History in American Life*, Roy and Thelen argued, in the teeth of pundits to Right and Left alike, that ordinary Americans make their own ways of finding contact with the past. They deftly dramatized the richness, interest, and intractability of the interviews that formed the core of their source materials. "On June 5, 1994," they write at the start of the book, "the past lay piled up on the porch of Roy's house in Arlington, Virginia. It lay on our laps in thick spiral-bound volumes and printouts of computer-generated tables as we tried to get a hold on what we had learned since March. How *do* Americans understand their pasts?" With characteristic imagination and passion, they drew on this material to show that Americans in fact regularly seek out the past—not the historian's Big Past of national narrative and debates about causation, but a personal past, one narrated by parents and relatives, experienced through collections of coins and crockery, traditional dances, holidays, and food, as much in grandparental kitchens as in museums—and one that shows considerable powers of imagination.

At the same time, both authors resisted any temptation to exaggerate the depth and richness of these connections. In his concluding reflections Roy characteristically suggested ways for bringing together professionals and amateurs, "Civil War reenactors and Civil War historians," "professional archivists" and "popular historymakers," and tapping into "the intimate ways that people use the past." He discussed public historians' efforts to create an ethic of shared authority, working with community members to shape museums and exhibits—and the unexpected and sometimes insuperable problems that arise when such ideals are put into practice. Roy warmly appreciated ordinary Americans' ability to see the past in moral and human terms and to infuse their visions of it with passion. But he remained convinced that these connections to the past needed reinforcement and enrichment. And he insisted that professionals who tried to feed the public's hunger for information about the past faced a task as difficult as it was vital: "providing context and comparisons and offering structural explanations" in a form that could appeal to, rather than alienate, ordinary readers. Part of Roy's legacy—perhaps the most important part—is this charge to historians of the generations to come.

The essays collected in this volume reflect Roy's vision and his achievements in many ways. But they shed a particularly bright light on his accomplishments as a devoted critic of Clio's new digital incarnation. He saw digital history as a set of tools rather than a panacea and made clear, again and again, that neither hymns of praise nor shrieks of fear did justice to the multiple ways they could be employed, as social, economic, and technical contexts determined:

> As with views of the Internet, in general, scholars should avoid unreflective boosterism or instinctive Luddism. New technologies could narrow access to students with the resources to purchase the latest equipment. And university administrators eager to cut costs or large corporations looking for new arenas for profitmaking could also use the new technology for their own ends. But at the same time, new technology opens up the resources of the Library of Congress to students at institutions without extensive libraries and it offers ways for scholars to create their own teaching materials without the mediation of giant publishing conglomerates. Neither the democratization or the commodification of higher education is inherent in the technology.

Many of the Web's other pioneers made wildly inflated claims for electronic media. Roy, by contrast, produced an ongoing critical commentary, informed, skeptical, and tolerant, on the transformation in the methods of the humanities that he himself was helping to bring about. In the 1990s and after, many claimed that distance learning could replace costly face-to-face classes with online learning that would be equal or superior in quality to the traditional kind and vastly cheaper to provide. "The evidence I have seen so far," Roy commented in 2006, "suggests that good quality distance education is more expensive and labor intensive than in-person instruction"—though he went on to make clear that it could certainly have its uses for students in remote locations and for those who could not travel. In the first heyday of hypertext, many claimed that the new forms of presentation made possible by the Web would soon render traditional articles and monographs obsolete. No one appreciated more clearly than Roy that the new electronic media enabled scholars to weave arguments together with the full body of evidence that supported them—and not only to use a vast range of visual and audio evidence but also to make possible a new kind of "interactivity" between users and these vivid, previously inaccessible sources. Yet Roy also noted that "the Web has not been designed with humanities scholarship in mind"—and therefore handled simple footnote references less well than word processors developed early in the history of the PC. More important, he never forgot that the same Web-based sources that might seem universally accessible to middle-class Americans might well be impossible for Latin Americans or Africans to use at all—or might embody a view of the past that looked just as parochial, to visitors from other societies and cultures, as the histories written by previous generations of American historians. To read Roy's articles in their historical context is to feel a new respect for his judgment, insight, and prescience, and to understand the digital humanities in a newly rounded, comprehensive way.

The past, Roy insisted, was not and is not dead. Digital media mattered to him because they offered a new and powerful way to keep the past alive and to make it rewarding and attractive. They could, when properly mobilized, bring the millions of Americans who care about the past into direct contact with primary sources and the lives and works that they record. The new devices that Roy mastered and created, the center that he built, the projects that he found money and expertise and ingenuity and energy to create—all were meant, in the end, to help ordinary men and women, old and young, to find the documents and images and songs that can give them a past rich with meaning and power. Not only every man, in Carl Becker's famous terms, but everyone might potentially be a historian—and find the necessary inspiration and information, freely available, on his or her computer screen. Yet realizing this promise, like any other, would require thought, devotion, and years of struggle.

A lifelong radical, Roy showed, by precept and example, that those ancient liberal ideals, humanism and scholarship, can flourish in the buzzing electronic landscape that we now inhabit—showed that they may, in the digital world, become richer, as well as more accessible, than they have ever been before. He never fell prey to the illusion that this would happen easily or soon. But he also never lost his faith or his commitment, and he fought in the open, generously and cheerfully, to realize his ideals. Thanks to this book, new readers will have a chance to share with Roy's friends the memory of a brave and brilliant historian who created small republics of letters wherever he went and always practiced his craft in the service of the democratic intellect.

Note to Readers

Deborah Kaplan

I N 2005, ROY ROSENZWEIG BEGAN PLANNING FOR A book of essays that would track significant developments in the field of digital history during its earliest years and consider possible directions for its future. He intended to include articles that he had already published and to write new ones. This is not the essay collection that Roy envisioned, but it is not very far from it. He was an early and enthusiastic adopter—of word processing, databases, and e-mail (although messages need recipients, and, in the beginning, there were few others online with whom he could try communicating)—and he first published an article about his experience with the new media in the early 1980s. But the pieces in this collection were written later, spanning the twelve-year period 1994 to 2006, when Roy became an incisive and influential commentator on digital history. Although I am aware that a chronological imperative often organizes books about history, I have selected and arranged the essays to emphasize three ways Roy engaged with the new technologies he loved so much.

Drawn to what the technology was able to do, he explored ways of practicing history that it made possible; much of his deep knowledge of new media came from projects he created with friends and colleagues at the American Social History Project, at the Center for History and New Media, and with professional organizations such as the Organization of American Historians. In the socially and intellectually rich last fifteen years of his life he collaborated with others on multimedia history CD-ROMs and Web sites that provide a wealth of primary and secondary sources for teachers and students as well as training in how to think analytically about them; on experiments in creating and disseminating scholarly research in new media; on digital tools for scholars, archivists, museum professionals, and history buffs; and on digital archives that gather eyewitness testimonies about important contemporary events. Forming one section of *Clio Wired*, the essays he wrote about practicing history in new media are a mix of reports from the field and policy papers. He also shared his experiences in a practical, how-to book that he wrote with CHNM colleague and friend Dan Cohen, *Digital History: A Guide to Gathering, Preserving, and Presenting the Past on the Web*, and I have included one of the chapters, "Collecting History Online," in the section. Fittingly, two of the other essays were also coauthored with friends with whom he had done hands-on work: Steve Brier and Randy Bass.

At the same time that he and his collaborators were teaching themselves to be digital historians and writing about what they were producing, Roy surveyed this quickly growing field, by publishing essays on the World Wide Web and on the history of the Internet itself. Composing another section of this book, those essays are surveys in that they move chronologically but also in the sense that they both map and carefully look and appraise. Roy assumed the roles of guide and evaluator in order to help others understand the dramatic changes taking place as history and its study were being translated into the new media. With another close friend and collaborator, Mike O'Malley, he wrote one of these essays, "Brave New World or Blind Alley? American History on the World Wide Web," and five years later he revisited and reassessed what had become a much more populous terrain in "The Road to Xanadu: Public and Private Pathways on the History Web."

Over time Roy came to feel that the revolutionary potential of new media—their tendency to disrupt conventional methods and aims in the discipline—was more important than the uses to which the technology was put. He observed, at the professional meetings and panels in which he participated, that a more narrow technical focus would routinely open out to "the most profound questions about why we do what we do as historians." Essays that he wrote on aspects of digital history, particularly between 2001 and 2006, reflect on such fundamental concerns. "Scarcity or Abundance: Preserving the Past," for example, considers what evidence of the past should be preserved and who should be responsible for preserving it. "Web of Lies? Historical Knowledge on the Internet," which he wrote with Dan Cohen, explores what makes historical knowledge authoritative on the history Web, how historians measure student learning, and why digital history might enable and require new research methods. What is the purpose of a scholarly article, and what differentiates amateur from professional history? These are two of the questions that Roy's "*Wikipedia*: Can History Be Open Source?" examines. I have put this group of essays first in the collection because they have the most far-reaching implications. In identifying and rethinking many taken-for-granted assumptions and practices, Roy provided not only a record of what has happened as the field of history has moved into new media but also broader assessments of what it has meant and will mean "to do history."

The categories of "Rethinking," "Practicing," and "Surveying" are admittedly overly neat. I hope that they serve to identify the basic orientations of Roy's essays even if few of the pieces fit comfortably under only one rubric. Some of his strongest convictions, evident even in his pre-digital history scholarship, run through these pieces and tend to highlight commonalities rather than differences among the essays. For example, he believed that academics needed, in effect, to get out more. He was an advocate for enlarging the view of historical study, which also means reenvisioning the community of those practicing history. Roy felt that university faculty should work more often alongside K-12 and community college teachers, as they all needed to develop and learn effective uses of online materials for teaching history. He urged historians to make common cause with librarians and archivists, to help in preserving the digital past, in collecting and making accessible historical sources online, and in influencing public policies guiding these efforts. He was equally interested in closer connections between academics and museum workers. Having written extensively about public and popular history before they made their way into digital media, Roy thought hard about what it means for museums to put their collections online, to make objects virtual. (He had hoped to write an essay for this book considering how important to museums the actual artifacts are, a compelling question for historians and curators alike.) And he appreciated the projects of history buffs who collect and present the past online. He showed that their work has much to teach academics, just as he thought that professional historians had skills that should be shared with amateurs.

Roy was committed, in his words, to "fostering a more democratic history." He understood the objective he sought as a *practice*—"a history that can be participated in by many different people, not just a detached, academic elite"—and as the *content* produced by that practice—"a history that incorporates the stories of ordinary people and is open to multiple voices and diverse perspectives." But it also meant to him wide *availability*: "a history that is accessible to multiple audiences." This last feature gave moral force and urgency to his writing. While his essays often express his admiration for digital media because they make it possible for people who would not otherwise have a way to do so to encounter historical resources, they also voice his opposition to restrictions on access, particularly those engendered by conglomerates wishing to turn a profit on the Internet.

He often concluded his essays with a call to action. Roy wanted to persuade historians to make their own writings and sources available for free on the Internet. He proposed that they contribute to collective projects that gather and present history online. He hoped that these efforts would inspire further innovation and change the form of professional history. But his calls for action meant more than doing innovative academic work, much as he valued it. Although he worried about the increasing privatization and commercialization of the history Web, he was convinced that its future was not yet determined and that it was up to historians to shape it. Protecting and strengthening the public Web, Roy believed wholeheartedly, also requires political engagement here, now.

CLIO WIRED

Rethinking History in New Media

Scarcity or Abundance?

Preserving the Past

ON OCTOBER 11, 2001, THE SATIRIC BERT IS EVIL! Web site, which displayed photographs of the furry Muppet in Zelig-like proximity to villains such as Adolf Hitler (see figure 1.1), disappeared from the Web—a bit of collateral damage from the September 11th attacks. Following the strange career of Bert Is Evil! shows us possible futures of the past in a digital era—futures that historians need to contemplate more carefully than they have done so far.

In 1996, Dino Ignacio, a twenty-two-year-old Filipino Web designer, created Bert Is Evil! ("brought to you by the letter H and the CIA"), which became a cult favorite among early tourists on the World Wide Web. Two years later, Bert Is Evil! won a "Webby" as the "best weird site." Fan and "mirror" sites appeared with some embellishing on the Bert Is Evil! theme. After the bombing of the U.S. embassies in Kenya and Tanzania in 1998, sites in the Netherlands and Canada paired Bert with Osama bin Laden.[1]

This image made a further global leap after September 11. When Mostafa Kamal, the production manager of a print shop in Dhaka, Bangladesh, needed some images of bin Laden for anti-American posters, he apparently entered the phrase "Osama bin Laden" in Google's image search engine. The Osama and Bert duo was among the top hits. *Sesame Street* being less popular

FIGURE 1.1 Bert the Muppet at Hitler's side, from the now-defunct Bert Is Evil! Web site.

in Bangladesh than in the Philippines, Kamal thought the picture a nice addition to an Osama collage. But when this transnational circuit of imagery made its way back to more *Sesame Street*–friendly parts of the world via a Reuters photo of anti-American demonstrators (see figure 1.2), a storm of indignation erupted. Children's Television Workshop, the show's producers, threatened legal action. On October 11, 2001, a nervous Ignacio pushed the delete key, imploring "all fans [*sic*] and mirror site hosts of 'Bert is Evil' to stop the spread of this site too."[2]

Ignacio's sudden deletion of Bert should capture our interest as historians since it dramatically illustrates the fragility of evidence in the digital era. If Ignacio had published his satire in a book or magazine, it would sit on thousands of library shelves rather than having a more fugitive existence as magnetic impulses on a Web server. Although some historians might object that the Bert Is Evil! Web site is of little historical significance, even traditional historians should worry about what the digital era might mean for the historical record. U.S. government records, for example, are being lost on a daily basis. Although most government agencies started using e-mail and word processing in the mid-1980s, the National Archives still does not require that digital records be retained in that form, and governmental employees profess confusion over whether they should be preserving electronic

FIGURE 1.2 Supporters of Osama bin Laden demonstrate against the United States in Dhaka, Bangladesh, on October 11, 2001. The poster, which was created from images downloaded from the Internet, includes bin Laden juxtaposed with the *Sesame Street* character Bert. Copyright Reuters 2001. Reuters News Picture Service Photo by Rafiqur Rahman.

files.[3] Future historians may be unable to ascertain not only whether Bert is evil but also which undersecretaries of defense were evil, or at least favored the concepts of the "evil empire" or the "axis of evil." Not only are ephemera like Bert and government records made vulnerable by digitization but so are traditional works—books, journals, and film—that are increasingly being born digitally. As yet, no one has figured out how to ensure that the digital present will be available to the future's historians.

But, as we shall see, tentative efforts are afoot to preserve our digital cultural heritage. If they succeed, historians will face a second, profound challenge—what would it be like to write history when faced by an essentially complete historical record? In fact, the Bert Is Evil! story could be used to tell a very different tale about the promiscuity and even persistence of digital materials. After all, despite Ignacio's pleas and Children's Television Workshop's threats, a number of Bert "mirror" sites persist. Even more remarkably, the Internet Archive—a private organization that began archiving the Web in 1996—has copies of Bert Is Evil! going back to March 30, 1997. To be sure, this extraordinary archive is considerably more fragile than one would like. The continued existence of the Internet Archive rests largely on the interest and energy of a single individual, and its collecting of copyrighted material is on even shakier legal ground. It has put the future of the past—traditionally seen as a public patrimony—in private hands.

Still, the astonishingly rapid accumulation of digital data—obvious to anyone who uses the Google search engine and gets 300,000 hits—should make us consider that future historians may face information overload. Digital information is mounting at a particularly daunting rate in science and government. Digital sky surveys, for example, access over 2 billion images. Even a dozen years ago, NASA already had 1.2 million magnetic tapes (many of them poorly maintained and documented) with space data. Similarly, the Clinton White House, by one estimate, churned out 6 million e-mail messages per year. And NARA is contemplating archiving military intelligence records that include more than "1 billion electronic messages, reports, cables, and memorandums."[4]

Thus historians need to be thinking simultaneously about how to research, write, and teach in a world of unheard-of historical abundance *and* how to avoid a future of record scarcity. Although these prospects have occasioned enormous commentary among librarians, archivists, and computer scientists, historians have almost entirely ignored them. In part, our detachment stems from the assumption that these are "technical" problems, which are outside the purview of scholars in the humanities and social sciences. Yet the more important and difficult issues about digital preservation are social, cultural, economic, political, and legal—issues that humanists should excel at. The "system" for preserving the past that has evolved over centuries is in crisis, and historians need to take a hand in building a new system for the coming century. Historians also tend to assume a professional division of responsibility, leaving these matters to archivists. But the split of archivists from historians is a relatively recent one. In the early twentieth century, historians saw themselves as having a responsibility for preserving as well as researching the past. At that time, the vision and membership of the American Historical Association—embracing archivists, local historians, and "amateurs" as well as university scholars—was considerably broader than it later became.[5]

Ironically, the disruption to historical practice (to what Thomas Kuhn called "normal science") brought by digital technology may lead us "back to the future." The struggle to incorporate the possibilities of new technology into the ancient practice of history has led, most importantly, to questioning the basic goals and methods of our craft. For example, the Internet has dramatically expanded and, hence, blurred our audiences. A scholarly journal is suddenly much more accessible to high school students and history enthusiasts. And the work of history buffs is similarly more visible and accessible to scholars. We are forced, as a result, to rethink who our audiences really are. Similarly, the capaciousness of digital media means that the page limits of scholarly journals are no longer fixed by paper and ink costs. As a result, we are led to question the nature and purpose of these journals—why do they publish articles with particular lengths and structures? Why do they publish particular types of articles? The simultaneous fragility and promiscuity of digital data requires yet more rethinking—about whether we should be trying to save everything, who is "responsible" for preserving the past, and how we find and define historical evidence.

Historians, in fact, may be facing a fundamental paradigm shift from a culture of scarcity to a culture of abundance. Not so long ago, we worried about the small numbers of people we could reach, pages of scholarship we could publish, primary sources we could introduce to our students, and documents that had survived from the past. At least potentially, digital technology has removed many of these limits: over the Internet, it costs no more to deliver the *American Historical Review* to 15 million people than 15,000 people; it costs less for our students to have access to literally millions of primary sources than a handful in a published anthology. And we may be able to both save and quickly search through all of the products of our culture. But will abundance bring better or more thoughtful history?[6]

Historians are not unaware of these challenges to the ways that we work. Yet, paradoxically, these fundamental questions are often relegated to more marginal professional spaces—to casual lunchtime conversations or brief articles in association newsletters. But in this time of rapid and perplexing changes, we need to engage with issues about access to scholarship, the nature of scholarship, the audience for scholarship, the sources for scholarship, and the nature of scholarly training in the central places where we practice our craft—scholarly journals, scholarly meetings, and graduate classrooms. That scholarly engagement should also lead us, I believe, to public action to advocate the preservation of the past as a *public* responsibility—one that historians share. But I hope to persuade even those who do not share my particular political stance that professional historians need to shift at least some of their attention from the past to the present and future and reclaim the broad professional vision that was more prevalent a century ago. The stakes are too profound for historians to ignore the future of the past.

Although historians have mostly been silent, archivists, librarians, public officials, and others have loudly warned about the threatened loss of digital records and publications for at least two decades. Words such as "disaster" and "crisis" echo through their reports and conference proceedings. As early as 1985, the Committee on the Records of Government declared, "the United States is in danger of losing its memory."[7] More than a dozen years later, a project called "Time and Bits: Managing Digital Continuity" brought together archivists, librarians, and computer scientists to address the problem once again. Conferees watched the Terry Sanders film *Into the Future: On the Preservation of Knowledge in the Electronic Age*, and some likened it to Rachel Carson's *Silent Spring* and themselves to the environmentalists of the 1960s and 1970s. A *Time and Bits* Web site assembled conference materials and promoted "ongoing digital dialogue." But, as if to prove the conference's point, the site disappeared in less than a year. Computer scientist Jeff Rothenberg may have been overoptimistic when he quipped, "Digital documents last forever—or five years, whichever comes first."[8]

Those worried about a problem like digital preservation that lacks public attention are prone to exaggerate. Probably the greatest distortion has been the implicit suggestion that we have somehow fallen from a golden age of preservation in which everything of importance was saved. But much—really, most—of the record of previous historical eras has disappeared. "The members of prehistoric societies did not think they lived in prehistoric times," *Washington Post* writer Joel Achenbach observes. "They merely lacked a good preservation medium." And nondigital records that have survived into this century—from Greek and Chinese antiquities to New Guinean folk traditions to Hollywood films—are also seriously threatened.[9]

Another exaggeration involves stories about the grievous losses that never occurred. One widely repeated story is that computers can no longer read the data tapes from the 1960 U.S. Census. In truth, as Margaret Adams and Thomas Brown from the National Archives have shown, the Census Bureau had by 1979 successfully copied almost all the records to newer "industry-compatible tapes." Yet, even in debunking one of the persistent myths of the digital age, Adams and Brown reveal some of the key problems. In just a decade and a half, migrating the census tapes to a readable format "represented a major engineering challenge"—hardly something we expect to face with historical records originating from within our own lifetimes. And although "only 1,575 records . . . could not be copied because of deterioration," the absolute nature of digital corrosion is sobering.[10] Print books and records decline slowly and unevenly—faded ink or a broken-off corner of a page. But digital records fail completely—a single damaged bit can render an entire document unreadable. Here is the key difference from the paper era: we need to take action now because digital items very quickly become unreadable or recoverable only at great expense.

This has already happened—albeit not as much as sometimes suggested. "Ten to twenty percent of vital data tapes from the Viking Mars mission," notes Deanna Marcum, the president of the Council on Library Information Resources, "have significant errors because magnetic tape is too susceptible to degradation to serve as an archival storage medium." Often, records lack sufficient information about their organization and coding to make them usable. According to Kenneth Thibodeau, director of the National Archive and Record Administration's Electronic Records Archives program, NARA lacked adequate documentation to make sense of several hundred reels of computer tapes from the Department of Health and Human Resources and data files from the National Commission on Marijuana and Drug Abuse. Some records could be recovered by future digital archaeologists, but sometimes only through an unaffordable "major engineering challenge."[11] The greatest concern is not over what has already been lost but what historians in fifty years may find that they can't read.

Many believe—incorrectly—the central problem to be that we are storing information on media with surprisingly short life spans. To be sure, acid-free paper and microfilm last a hundred to five hundred years, whereas digital and magnetic media deteriorate in ten to thirty years. But the medium is far from the weakest link in the digital preservation chain. Well before most digital media degrade, they are likely to become unreadable because of changes in hardware (the disk or tape drives become obsolete) or software (the data are organized in a format destined for an application program that no longer works). The life expectancy of digital media may be as little as ten years, but very few hardware platforms or software programs last that long. Indeed, Microsoft only supports its software for about five years.[12]

The most vexing problems of digital media are the flipside of their greatest virtues. Because digital data are in the simple lingua franca of bits, of ones and zeros, they can be embodied in magnetic impulses that require almost no physical space, be transmitted over long distances, and represent very different objects (for instance, words, pictures, or sounds as well as text). But the ones and zeros lack intrinsic meaning without software and hardware, which constantly change because of technological innovation and competitive market forces. Thus this lingua franca requires translators in every computer application, which, in turn, operate only on specific hardware platforms. Compounding the difficulty is that the languages being translated keep changing every few years.

The problem is still worse because of the ability of digital media to create and represent complex, dynamic, and interactive objects—another of their great virtues. Even relatively simple documents that appear to have direct print analogs turn out to be more complex. Printing out e-mail messages makes rapid searches of them impossible and often jettisons crucial links to related messages and attachments. In addition, multimedia programs, which generally rely on complicated combinations of hardware and software, quickly become obsolete. Nor is there any good way to preserve interactive and experiential digital creations. That is most obviously true of computer games and digital art, but even a large number of ordinary Web pages are generated out of databases, which means that the specific page you view is your own "creation" and the system can create an infinite number of pages. Preserving hypertextually linked Web pages poses the further problem that to save a single page in its full complexity could ultimately require you to preserve the entire Web, because virtually every Web page is linked to every other. And the dynamic nature of databases destabilizes mundane business and governmental records since they are often embedded in systems that automatically replace old data with new—a changeability that, notes archival educator Richard Cox, threatens "the records of any modern day politician, civic leader, businessperson, military officer, or leader."[13]

While these technical difficulties are immense, the social, economic, legal, and organizational problems are worse. Digital documents—precisely because they are in a new medium—have disrupted long-evolved systems of trust and authenticity, ownership, and preservation. Reestablishing those systems or inventing new ones is more difficult than coming up with a long-lived storage mechanism.

How, for example, do we ensure the "authenticity" of preserved digital information and "trust" in the repository? Paper documents and records also face questions about authenticity, and forgeries are hardly unknown in traditional archives. The science of "diplomatics," in fact, emerged in the seventeenth century as a way to authenticate documents when scholars confronted rampant forgeries in medieval documents. But digital information—because it is so easily altered and copied, lacks physical marks of its origins, and, indeed, even the clear notion of an "original"—cannot be authenticated as physical documents and objects can. We have, for example, no way of knowing that forwarded e-mail messages we receive daily have not been altered. In fact, the public archive of Usenet discussion groups contains hundreds of deliberately and falsely attributed messages. "Fakery," write David Bearman and Jennifer Trant, "has not been a major issue for most researchers in the past, both because of the technical barriers to making plausible forgeries, and because of the difficulty with which such fakes entered an authoritative information stream."[14] Digital media, tools, and networks have altered the balance.

"It took centuries for users of print materials to develop the web of trust that now undergirds our current system of publication, dissemination, and preservation," notes Abby Smith, a leading figure in library and preservation circles. Digital documents are disrupting that carefully wrought system by undercutting our expectations of what constitutes a trusted and authentic document and repository. But to make the transition to a new system requires not just technical measures (such as digital signatures and "watermarks") but, as Clifford Lynch, the executive director of the Coalition for Networked Information, observes, also figuring out responsibility for guaranteeing claims of authorship and financing for a system of "authentication and integrity management."[15]

Such questions are particularly hard to answer since digitization also undercuts our sense of who owns such materials and, thus, who has the right and responsibility to preserve them. Consumers (including libraries) have traditionally purchased books and magazines under the "first sale" doctrine, which gives those who buy something the right to make any use of it, including lending or selling it to others. But most digital goods are licensed rather than sold. Because contract law governs licenses, vendors of digital content can set any restrictions they choose—they can say that the contents may not be copied or cannot be viewed by more than one person at a time. Adobe's eBook reader even includes a warning that a book may not be read aloud.[16]

But if libraries don't *own* digital content, how can they preserve it? The problem will become even worse if publishers widely adopt copy protection schemes, as they are seriously considering doing for electronic books. Even a library that had the legal right to preserve the content would have no reason to assume that it would be able to do so; meanwhile, the publisher would have little incentive to keep the protection system functioning in a new software environment. In general, digital rights management systems and other forms of "trusted computing" undercut preservation efforts by embedding centralized control in proprietary systems. "If Microsoft, or the U.S. government, does not like what you said in a document you wrote," speculates Free Software advocate Richard Stallman, "they could post new instructions telling all computers to refuse to let anyone read that document."[17]

Licensed and centrally controlled digital content not only erodes the ability of libraries to preserve the past, it also undercuts their responsibility. Why should a library worry about the long-term preservation of something it does not own? But then, who will? Publishers have not traditionally assumed preservation responsibility since there is no obvious profit to be made in ensuring that something will be available or readable in a hundred years when it is in the public domain and can't be sold or licensed.[18]

The digital era has not only unsettled questions of ownership and preservation for traditional copyrighted material, it has also introduced a new, vast category of what could be called semipublished works, which lack a clear preservation path. The free content available on the Web is protected by copyright even though it has not been formally registered with the Library of Congress Copyright Office or sold by a publisher. That means that a library that decided to save a collection of Web pages—say, those posted by abortion rights organizations—would technically be violating copyright.[19] The absence of this "process" is the most fundamental problem facing digital preservation. Over centuries, a complex (and imperfect) system for preserving the past has emerged. Digitization has unsettled that system of responsibility for preservation, and an alternative system has not yet emerged. In the meantime, cultural and historical objects are being permanently lost.

Four different systems generally preserve cultural and historical documents and objects. Research libraries take responsibility for books, magazines, and other published cultural works, including moving images and recorded sound. Government records fall under the jurisdiction of the National Archives and a network of state and local archives.[20] Systems for maintaining other cultural and historical materials are less formal or centralized. "Records" and "papers" from businesses, voluntary associations, and individuals have found their way into local historical societies, specialized archives, and university special collections. Finally, the semipublished body of material we have called "ephemera" has been most often saved by enthusiastic individuals—for example, postcard and comic book collectors—who might later deposit their hoard in a permanent repository.[21]

While research libraries have tried to save relatively complete sets of published works, other historical sources have generally only been preserved in a highly selective and sometimes capricious fashion—what archivists call "preservation through neglect." Materials that lasted fifty or one hundred years found their way into an archive, library, or museum. Although this inexact system has resulted in many grievous losses to the historical record, it has also given us many rich collections or personal and organizational papers and ephemera.[22]

But this "system" will not work in the digital era because preservation cannot begin twenty-five years after the fact. What might happen, for example, to the records of a writer active in the 1980s who dies in 2003 after a long illness? Her heirs will find a pile of unreadable 5¼″ floppy disks with copies of letters and poems written in WordStar for the CP/M operating system or one of the more than fifty now-forgotten word-processing programs used in the late 1980s.[23]

Government archives similarly continue to rely on the unwarranted assumption that records can be appraised and accessioned many years after their creation. A recent study, "Current Recordkeeping Practices Within the Federal Government," which surveyed more than forty federal agencies, found widespread confusion about "policies and procedures for managing, storing, and disposing of electronic records and systems." "Government employees," it concluded, "do not know how to solve the problem of electronic records—whether the electronic information they create constitutes records and, if so, what to do with the records. Electronic files that qualify as records—particularly in the form of e-mail, and also word processing and spreadsheet documents—are not being kept at all as records in many cases."[24]

This uncertainty and disarray would not be so serious if we could assume that it could be simply sorted out in another thirty years. But if we hope to preserve the present for the future, then the technical problems facing digital preservation as well as the social and political questions about authenticity, ownership, and preservation policy need to be confronted now.

At least initially, archivists and librarians tended to assume that a technical change—the rise of digital media—required a technical solution. The simplest technical solution has been to translate digital information into something more familiar and reassuring like paper or microfilm. But, as Rothenberg points out, this is a "rear-guard action" that destroys "unique functionality (such as dynamic interaction, nonlinearity, and integration)" and "core digital attributes (perfect copying, access, distribution, and so forth)" and sacrifices the "original form, which may be of unique historical, contextual, or evidential interest."[25]

Another backward-looking solution is to preserve the original equipment. If you have files created on an Apple II, then why not keep one in case you need it? Well, sooner or later, a disk drive breaks or a chip fails, and unless you have a computer junkyard handy and a talent for computer repair, you are out of luck. "Technological preservation," moreover, requires intervention before it is too late to save not just the files but also the original equipment. The same can be said of what is probably the most widely accepted current method of digital preservation—"data migration," or moving the documents from a medium, format, or computer technology that is becoming obsolete to one that is becoming more common.[26] When the National Archives saved the 1960 U.S. Census tapes, they used migration, and large organizations use this strategy all the time—moving from one accounting system to another. Because we have lots of experience migrating data, we also know that it is time consuming and expensive. One estimate is that data migration is equivalent to photocopying all the books in a library every five years.[27]

Some like Rothenberg also worry, for example, about the loss of functionality in migrating digital files. Moreover, the process can't be automated because "migration requires a unique new solution for each new format or paradigm and each type of document that is to be converted into that new form." Rothenberg is also derisive about the practice of translating documents into standardized formats and then retranslating as new formats emerge, which he finds "analogous to translating Homer into modern English by way of every intervening language that has existed during the past 2,500 years."[28]

Rothenberg's favored alternative is "emulation"—developing a system that works on later generations of hardware and software but mimics the original. In principle, a single emulation solution could preserve a vast store of digital documents. In addition, it holds the greatest promise for preserving interactive and multimedia digital creations. But critics of emulation tellingly note that it is only a theoretical solution. Probably the best strategy is to reject the all-or-nothing, magic-bullet approaches implicit in the proposals of the most passionate advocates of any particular strategy—whether creating hard copies, preserving old equipment, migrating formats, or emulating hardware and software. Margaret Hedstrom, one of the leading figures in digital preservation research, argues persuasively that "the search for the Holy Grail of digital archiving is premature, unrealistic, and possibly counter-productive." Instead, we need to develop "solutions that are appropriate, effective, affordable and acceptable to different classes of digital objects that live in different technological and organizational contexts."[29]

But even the most calibrated mix of technical solutions will not save the past for the future because, as we have seen, the problems are much more than technical and involve difficult social, political, and organizational questions of authenticity, ownership, and responsibility. Multiple experiments and practices are under way—more than can be discussed here. But I want to focus on some widely discussed approaches or experiments as illustrative of some of the possibilities and continuing problems.

One of the earliest and most influential approaches to digital preservation (and digital authenticity) was what archivists call the "Pitt Project," a three-year (1993–1996) research effort funded by the National Historical Publications and Records Commission (NHPRC) and centered at the University of Pittsburgh School of Information and Library Studies. For historians, what is most interesting (and sometimes puzzling) about the Pitt Project approach is the way that it simultaneously narrows and broadens the role of archives and archivists through its focus on "records as evidence" rather than "information." "Records," David Bearman and Jennifer Trant explain, "are that which was created in the conduct of business" and provide "evidence of transactions." Data or information, by contrast, Bearman "dismisses as nonarchival and unworthy of the archivist's attention."[30] From this point of view, the government's record of your Social Security account is vital but not the "information" contained in letters that you and others might have written complaining about the idea of privatizing Social Security.

The Pitt Project produced a pathbreaking set of "functional requirements for evidence in electronic record keeping"—in effect, strategies and tactics to ensure that electronic records produce legally or organizationally acceptable evidence of their transactions. Such a focus responds particularly well to worries about the "authenticity" of electronic records. But for historians (and for some archivists), the focus on records as evidence rather than records as sources of information, history, or memory seems disappointingly narrow. Moreover, as Canadian archivist Terry Cook points out, the emphasis on "redesigning computer systems' functional requirements to preserve the integrity and reliability of records" and assigning "long-term custodial control . . . to the creator of archival records" privileges "the powerful, relatively stable, and continuing creators of records capable of such reengineering" and ignores artists, activists, and "marginalized and weaker members of society" who have neither the resources nor inclination to produce "business acceptable communications."[31]

While the Pitt Project emphasizes archival professionalism, a narrowing of the definition of recordkeeping, a rejection of the custodial tradition in archives, and planning for more careful collecting in the future rather than action in the present, the Internet Archive has taken precisely the opposite approach. It represents a grassroots, immediate, enthusiast response to the crisis of digital preservation that both expands and further centralizes archival responsibility in ways that were previously unimaginable. Starting in September 1996, Brewster Kahle and a small staff sent "crawlers" out to capture the Web by moving link-by-link and completing a full snapshot every two months. Although in part a philanthropic venture funded by Kahle, the Internet Archive also has a commercial side. Kahle's for-profit Web navigation service, Alexa Internet (bought by Amazon in 1999 for $300 million), is what actually gathers the Web snapshots, which it uses to analyze patterns of Web use, and then donates them to the Internet Archive.[32]

By February 2002, the Internet Archive (IA) had gathered a monumental collection of more than 100 terabytes of Web data—about 10 billion Web pages or five times all the books in the Library of Congress—and was gobbling up 12 terabytes more each month. That same fall, it began offering public access to most of the collection through what Kahle called the "Wayback Machine"—a wry reference to the device used by the time-traveling Mr. Peabody in the Rocky and Bullwinkle cartoons of the 1960s. Astonishingly, a single individual with a very small staff has created the world's largest database and library in just five years.[33]

In December 2001, shortly after the Wayback Machine became public, the search engine company Google unveiled "Google Groups," another massive digital archive—this one under purely commercial auspices. Google Groups provides access to more than 650 million messages posted over the past two decades to "Usenet," the online discussion forums that pre-date even the Internet. Although "ownership" seems like a dubious concept in relation to a public discussion forum, Google purchased the archive from Deja.com, which had brought the groups to the Web but then collapsed in the Internet bust. Despite Deja.com's failure, Google sees the Usenet Archive as another attractive feature in its stable of online information resources and tools.[34]

Both IA and Google Groups are libraries organized on principles that are more familiar to computer scientists than to librarians, as Peter Lyman, who knows both worlds as the head of the University of California at Berkeley library and as a member of the IA board, points out. The library community has focused on developing "sophisticated cataloging strategies." But computer scientists, including Kahle, have been more interested in developing sophisticated search engines that operate directly on the data we see (the Web pages) rather than on the metadata (the cataloging information). Whereas archival and library projects focus on "high-quality collections built around select themes" and make the unit of cataloging the Web page, the computer science paradigm "allows for archiving the entire Web as it changes over time, then uses search engines to retrieve the necessary information."[35]

Projects designed by librarians and archivists generally have the advantages of precision and standardization. They favor careful protocols and standards such as the Dublin Core, the OAIS (Open Archival Information System), and the EAD (Encoded Archival Description). But the expense and difficulty of the protocols and procedures mean that less well funded and staffed archives and libraries often ignore them. Responding to presentations by advocates of standards at a conference, computer scientist Jim Miller warned that if archivists push for too much cataloging metadata "they might end up with none."[36]

The Internet Archive, which is the child of the search engines and the computer scientists, is an extraordinarily valuable resource. Most historians will not be interested now, but in twenty-five or fifty years they will delight in searching it. A typical college history assignment in 2050 might be to compare Web depictions of Muslim Americans in 1998 and 2008. But any appreciation of the IA must acknowledge its limitations. For example, large numbers of Web pages do not exist as "static" HTML pages; rather, they are stored in databases, and the pages are generated "on the fly" by search queries. As a result, the IA's crawlers do not capture much of the so-called "deep Web" that is stored in databases. Multimedia files—streaming media and flash—also do not seem to be captured. In addition, the Internet Archive's crawls cannot go on forever; at some point, they stop, since, as one of the computer scientists who manages them acknowledges, "the Web is essentially infinite in size." Anyone who browses the IA regularly encounters such messages as "Not in Archive" and "File Location Error" or even "closed for maintenance."[37]

Some pages are missing for legal and economic as well as technical reasons. Private, gated sites are off-limits to the Internet Archive's crawlers. And many ungated sites also discourage the crawlers. The *New York Times* allows free access to its current contents, but charges for articles more than one week old. If the IA gathered up and preserved the *Times*'s content, there would be no reason for anyone to pay the *Times* for access to its proprietary archive. As a result, the *Times* includes a "robots exclusion" file on its site, which the IA respects. Even those sites without the robots exclusion file and without any formal copyright are still covered by copyright law and could challenge the IA's archiving of their content. To avoid trouble, the IA simply purges the pages of anyone who complains. It is as if Julie Nixon could write to the National Archives and tell them to delete her father's tapes or an author could withdraw an early novel from circulation.[38]

Thus the Internet Archive is very far from the complete solution to the problem of digital preservation. It does not deal with the digital records that vex the National Archives and other repositories because they lack the public accessibility and minimal standardization in HTML of Web pages. Nor does it include much formally published literature—e-books and journals—which is sold and hence gated from view. And even for what it has gathered, it has not yet hatched a long-term preservation plan, which would have to incorporate a strategy for continuing access to digital data that are in particular (and time-bound) formats. Even more troubling, it has no plan for how it will sustain itself into the future. Will Kahle continue to fund it indefinitely?[39] What if Amazon and Alexa no longer find it worthwhile to gather the data, especially since acquisition costs are doubling every year?

Similar questions could be raised about "Google Groups." What if the company decides that there is no prospect of gaining adequate advertising revenue by making old newsgroup messages available (as, indeed, Deja.com previously determined)? While appreciating Google's entrepreneurial energy in preserving and making available an enormous body of historical documents, we should also look carefully at the way private corporations have suddenly entered into a realm—archives—that was previously part of the public sector—a reflection of the privatization sweeping across the global economy. At least so far, our most important, and most imaginatively constructed, digital collections are in private hands.[40]

Given that the preservation of cultural heritage and national history are arguably social goods, why shouldn't the government take the lead in such efforts? One reason is that at least some key aspects of the digital present—the Bert story, for example—do not follow national boundaries and, indeed, erode them. If national archives were part of the projects of state-building and nationalism, then why should states support postnational digital archives? The declining significance of state-based national archives may mirror the decline of the contemporary national state. So far, the Smithsonian Institution and the Library of Congress have worked with the Internet Archive only where they needed its help in documenting some particularly national stories—the elections of 1996 and 2000 and the September 11th attacks.

Another reason for the limited government role is that the digital preservation crisis emerged most dramatically during the antistatist Reagan revolution of the 1980s. In the 1970s, for example, the electronic records program of the National Archives made a modest, promising start. But, as archivist Thomas E. Brown writes, it went into "a near total collapse in the 1980s." The staff dropped to seven people by 1983, and, amazingly, this beleaguered group charged with guarding the nation's electronic records had no access to computer facilities. Things began to improve in the early 1990s, but, after 1993, the electronic records program suffered from further cutbacks in the federal workforce. An underfunded and understaffed National Archives was hardly in a position to develop a solution to the daunting and mounting problem of electronic federal records.[41]

The Library of Congress also initially eschewed a leading role in preserving digital materials, as the National Research Council later complained. Here, too, one could detect the weakening influence of the state. The library's high-profile effort in the digital realm was "American Memory," which digitized millions of items from its collections and placed them online. Teachers, students, and researchers love American Memory, but it did nothing to preserve the growing number of "born digital" objects. Not coincidentally, American Memory was a project that could attract large numbers of private and corporate donors, who often saw sponsorship as good advertising and who paid for three-quarters of the project.[42]

Better developed state-centered approaches to digital preservation have, not surprisingly, emerged outside the United States—in Australia and Scandinavia, for example. Norway requires that digital materials be legally deposited with the national library in return for copyright protection.[43] One of the key ways that the Library of Congress could help preserve the future of digital materials would be to aggressively assert its copyright deposit claims, which would finesse some of the legal and ownership issues troubling the Internet Archive.[44]

Nevertheless, the National Archives and the Library of Congress have very recently begun—prodded by outside critics and supported belatedly by Congress—to take a more aggressive approach on digital preservation. The archives is proposing a "Redesign of Federal Records Management" to respond to the reality that "a large majority of electronic record series of continuing value are not coming into archival custody." It is also working closely with the San Diego Supercomputing Center on developing "persistent object preservation" (POP), which creates a description of a digital object (and groups of digital objects) in simple tags and schemas that will be understandable in the future; the records would be "self-describing" and, hence, independent of specific hardware and software. The computer scientists maintain that records in this format will last for three hundred to four hundred years.[45]

In December 2000, the Library of Congress launched the most important initiative, the National Digital Information Infrastructure Program (NDIIP). Even this massive and important federal initiative bore the marks of the anti-statist, privatization politics of the 1980s. Congress gave the library $5 million for planning and promised another $20 million when it approved the plan. But the final $75 million will only be distributed as a match against an equal amount in private funds.[46]

Although the future of the digital present remains perilous, these recent initiatives suggest some encouraging strategies for preserving the range of digital materials. A combination of technical and organizational approaches promises the greatest chance of success, but privatization poses grave dangers for the future of the past. Advocates of digital preservation need to mobilize state funding and state power (such as the assertion of eminent domain over copyright materials) but infuse it with the experimental and ad hoc spirit of the Internet Archive. And we need to recognize that, for many digital materials (especially the Web), the imperfect computer-science paradigm probably has more to recommend it than the more careful and systematic approach of the librarians and archivists. What is often said of military strategy seems to apply to digital preservation: "the greatest enemy of a good plan is the dream of a perfect plan."[47] We have never preserved everything; we need to start preserving something.

Given the enormous barriers to saving digital records and information, it comes as something of a surprise that many continue to insist that a perfect plan—or at least a pretty good plan—will eventually emerge. Techno-optimists such as Brewster Kahle dream most vividly of the perfect plan and its startling consequences. "For the second time in history," Kahle writes with two collaborators, "people are laying plans to collect all information—the first time involved the Greeks which culminated in the Library of Alexandria. . . . Now . . . many [are] once again to take steps in building libraries that hold complete collections." Digital technology, they explain, has "gotten to the point where scanning all books, digitizing all audio recordings, downloading all Websites, and recording the output of all TV and radio stations is not only feasible but less costly than buying and storing the physical versions."[48] Librarians and archivists remain skeptical of such predictions, pointing out the enormous costs of cataloging and making available what has been preserved, and that we have never saved more than a fraction of our cultural output. But, whatever our degree of skepticism, it is still worth thinking seriously about what a world in which everything was saved might look like.

Most obviously, archives, libraries, and other record repositories would suddenly be freed from the tyranny of shelf space that has always shadowed their work. Digitization also removes other long-term scourges of historical memory such as fire and war. The 1921 fire that destroyed the 1890 census records provided a crucial spark that finally led to the creation of the National Archives. But what if there had been multiple copies of the census? The ease—almost inevitability—of the copying of digital files means that it is considerably less likely today that things exist in only a single copy.[49]

What would a new, virtual, and universal Alexandria library look like? Kahle and his colleagues have forcefully articulated an expansive democratic vision of a past that includes all voices and is open to all. "There are about ten to fifteen million people's voices evident on the Web," he told a reporter. "The Net is a people's medium: the good, the bad and the ugly. The interesting, the picayune and the profane. It's all there." Advocates of the new universal library and archive wax even more eloquently about democratizing access to the historical record. "The opportunity of our time is to offer universal access to all of human knowledge," said Kahle.[50]

Kahle's vision of cultural and historical abundance merges the traditional democratic vision of the public library with the resources of the research library and the national archive. Previously, few had the opportunity to come to Washington to watch early Thomas Edison films at the Library of Congress. And the library could not have served them if they had. Democratized access is the real payoff in electronic records and materials. It may be harder to preserve and organize digital materials than it is paper records, but, once that is accomplished, they can be made accessible to vastly greater numbers of people. To open up the archives and libraries in this way democratizes historical work. Already, people who had never had direct access to archives and libraries can now enter. High school students are suddenly doing primary source research; genealogy has exploded in popularity because you no longer have to travel to distant archives.

This vision of democratic access also promises direct and unmediated access to the past. Electronic commerce enthusiasts tout "disintermediation"—which is the elimination of the insurance and real estate broker and other intermediaries—and the emergence of one like eBay made up of only buyers and sellers. In theory, the universal digital library might bring a similar cultural disintermediation in which people interested in history make direct contact with the documents and artifacts of the past without the mediation of cultural brokers like librarians, archivists, and historians. Sociologist Mike Featherstone speculates on the emergence of a "new culture of memory" in which the existing "hierarchical controls" over access would disappear. This "direct access to cultural records and resources from those outside cultural institutions" could "lead to a decline in intellectual and academic power" in which the historian, for example, no longer stands between people and their pasts.[51] The "Wayback Machine" encapsulates this vision of disintermediation by suggesting that everyone, like Mr. Peabody and his boy Sherman, can jump in a time machine and find out what Columbus or Edison was "really" like. Of course, most historians would argue that untrained people are not likely to know how to think about what they find in digitized archives. Still, the balance of power may shift. Ask any travel agent how the widespread access to information undercuts professional control.

Most historians have not embraced this vision in which everyone becomes his or her own historian. Nor have they enthusiastically endorsed the vision of a universal library that contains all voices and all records. In my informal polling, most historians recoil at the thought that they would need to write history with even more sources.[52] Historians are not particularly hostile to new technology, but they are not ready to welcome fundamental changes to their cultural position or their modes of work. Having lived our professional careers in a culture of scarcity, historians find that a world of abundance can be unsettling.

Abundance, after all, can be overwhelming. How do we find the forest when there are so many damned trees? Psychologist Aleksandr Luria made this point in his famous study of a Russian journalist, "S" (S. V. Shereshevskii), who had an amazingly photographic memory; he could reproduce complex tables of numbers and long lists of words that had been shown to him years earlier. But this "gift" turned out to be a curse. He could not recognize people because he remembered their faces so precisely; a slightly different expression would register as a different person. Grasping the larger point of a passage or abstract idea "became a tortuous . . . struggle against images that kept rising to the surface in his mind." He lacked, as psychologist Jerome Bruner notes, "the capacity to convert encounters with the particular into instances of the general."[53]

If historians are to set themselves "against forgetting" (in Milan Kundera's resonant phrase), then they may need to figure out new ways to sort their way through the potentially overwhelming digital record of the past. Contemporary historians are already groaning under the weight of their sources. Robert Caro has spent twenty-six years working his way through just the documents on Lyndon B. Johnson's pre-vice-presidential years—including 2,082 boxes of Senate papers. Surely, the injunction of traditional historians to look at "everything" cannot survive in a digital era in which "everything" has survived.[54]

The historical narratives that future historians write may not actually look much different from those that are crafted today, but the methodologies they use may need to change radically. If we have, for example, a complete record of everything said in 2010, can we offer generalizations about the nature of discourse on a topic simply by "reading around"? Wouldn't we need to engage in some more methodical sampling in the manner of, say, sociology? Would this revive the social-scientific approaches with which historians flirted briefly in the 1970s? Wouldn't historians need to learn to write complex searches and algorithms that would allow them to sort through this overwhelming record in creative, but systematic, ways? The future gurus of historical research methodology may be the computer scientists at Google who have figured out how to search the equivalent of a 100-mile-high pile of paper in half a second. "To be able to find things with high accuracy and high reliability has an incredible impact on the world"—and, one might add, future historians. Future graduate programs will probably have to teach such social-scientific and quantitative methods as well as such other skills as "digital archaeology" (the ability to "read" arcane computer formats), "digital diplomatics" (the modern version of the old science of authenticating documents), and data mining (the ability to find the historical needle in the digital hay).[55] In the coming years, "contemporary historians" may need more specialized research and "language" skills than medievalists do.

Historians have time to think about changing their methods to meet the challenge of a cornucopia of historical sources. But they need to act more immediately on preserving the digital present or that reconsideration will be moot; they will be struggling with a scarcity, not an overabundance, of sources. Surprisingly, however, historians themselves have been scarce on this issue.[56] Archivists and librarians have intensely debated and discussed digitization and digital presentation for more than a decade. They have written hundreds of articles and reports, undertaken research projects, and organized conferences and workshops. Academic and teaching historians have taken almost no part in these conferences and have contributed almost nothing to this burgeoning literature. Historical journals have published nothing on the topic.[57]

Part of the reason is that preserving the born-digital materials for future historians seems like a theoretical and technical issue, tomorrow's problem or at least someone else's problem. Another reason for this disinterest is the divorce of archival concerns from the historical profession—a part of the general narrowing of the concerns of professional historians over the past century. In the late nineteenth and early twentieth centuries, historians and archivists were closely aligned. Perhaps the most important committee of the American Historical Association in the 1890s was the Historical Manuscripts Commission, which led to the AHA's influential Public Archives Commission. Archival concerns found a regular place in the AHA's Annual Meeting, the *American Historical Review*, and especially the voluminous AHA annual reports. Most important, the AHA led the fight to establish the National Archives. But in 1936 (in the midst of an earlier technological upheaval that came with the emergence of microfilm), the Conference of Archivists left the AHA to create the Society of American Archivists. The professions charged with writing about the past and preserving the records of the past have sharply diverged in the past seven decades. Today, only 82 of the 14,000 members of the AHA identify themselves as archivists.[58]

But historians ignore the future of digital data at their own peril. What, for example, about the long-term preservation of scholarship that is—increasingly—originating in digital form? Not only do historians need to ensure the future of their own scholarship, but linking directly from footnotes to electronic texts—an exciting prospect for scholars—will only be possible if a stable archiving system emerges.[59] For the foreseeable future, librarians and archivists will be making decisions about priorities in digital preservation. Historians should be at the table when those decisions are made. Do they wish to endorse, for example, the Pitt Project's emphasis on preserving records of business transactions rather than "information" more broadly?

One of the most vexing and interesting features of the digital era is the way that it unsettles traditional arrangements and forces us to ask basic questions that have been there all along. Some are about the relationship between historians and archival work. Should the work of collecting, organizing, editing, and preserving of primary sources receive the same kind of recognition and respect that it did in earlier days of the profession? Others are about whose overall responsibility it is to preserve the past. For example, should the National Archives expand its role in preservation beyond official records? For many years, historians have taken a hands-off approach to archival questions. With the unsettling of the status quo, they should move back more actively into this realm. If the Web page is the unit of analysis for the digital librarian and the link the unit of analysis for the computer scientists, what is the appropriate unit of analysis for historians? What would a digital archival system designed by historians look like? And how might we alter and enhance our methodologies in a digital realm? For example, in a world where all sources were digitized and universally accessible, arguments could be more rigorously tested. Currently, many arguments lack such scrutiny because so few scholars have access to the original sources—a problem that has arisen especially sharply in the recent controversies over Michael A. Bellesiles' *Arming America: The Origins of a National Gun Culture* (2000). In a new digital world, would historians then be held to the same standard of "reproducible" results as scientists?[60]

Of course, when historians get to the preservation table, they will discover a cultural and professional clash between their own impulses, which are to save everything, and those of librarians and archivists who believe that selection, whether passive or active, is inevitable. The National Archives, for example, only permanently accessions 2 percent of government records.[61] This conflict surfaced in the 1980s and 1990s, when librarians tried to bring in scholars to discuss priorities in preserving books that were deteriorating because of acidic paper. Librarians found the discussion "frustrating." "Many scholars," recalls Deanna Marcum, declared "that everything had to be saved and they could not make choices." Not surprisingly, scholars have responded very differently to Nicholson Baker's sharp attack on the microfilming and disposal of aging books and newspapers in *Double Fold* than have archivists and librarians. Whereas many scholars have shared Baker's outrage that books and newspapers have been destroyed, archivists and librarians have responded in outrage to what they see as his failure to understand the pressures that make it impossible to save everything. Whereas historians with their gaze fixed on the past worry about information scarcity (the missing letter or diary), archivists and librarians recognize that we now live in a world of over-whelming information abundance.[62] If historians are going to join in preser-vation discussions, they will have to make themselves better informed about the simultaneous abundance of historical sources and scarcity of financial re-sources that lead archivists and librarians to respond with exasperation to scholars' blithe insistence that everything must be saved.

Preservation of the past is, in the end, often a matter of allocating ade-quate resources. Perhaps the largest problem facing the preservation of elec-tronic government records has nothing to do with technology; it is, as vari-ous reports have noted, "the low priority traditionally given to federal records management." In the absence of new resources, the costs of preservation will come from the money that our society, in the aggregate, allocates for history and culture. Richard Cox, for example, has argued that a greater portion of the budget of the National Historical Publications and Records Commission (NHPRC) should go to electronic records preservation and management and correspondingly less money should go to the letterpress Documentary Editions that the commission also funds, since "most of the records repre-sented by the documentary editions are not immediately threatened." This stance does not endear him to documentary editors, who are much better represented among professional historians than are archivists.[63]

The alternative to squabbling over inadequate resources that are appropri-ated for these purposes is joint action to secure further funds. When Shirley Baker, president of the Association of Research Libraries, challenged histo-rian Robert Darnton's favorable review of Baker's book and noted "choices have always had to be made" in the absence of "greater public commitment to the preservation of the historical record," Darnton responded by urging the establishment of "a new kind of national library dedicated to the preser-vation of cultural artifacts" (including disappearing digital records) and funded by income generated by the sale or rental of bandwidth.[64] Such state-based solutions return us to the kind of alliance between historians and archi-vists that led to building of the National Archives in the 1930s, an era of growing rather than waning confidence in the nation-state. Historians need to join in lobbying actively for adequate funding for both current historical work and preservation of future resources. They should also argue forcefully for the democratized access to the historical record that digital media make possible. And they must add their voices to those calling for expanding copy-right deposit—and opposing copyright extension, for that matter—of digital materials so as to remove some of the legal clouds hanging over efforts like the Internet Archive and to halt the ongoing privatization of historical resources. Even in the absence of state action, historians should take steps individually and within their professional organizations to embrace the culture of abundance made possible by digital media and expand the public space of scholarship—for example, making their own work available for free on the Web, cross-referencing other digital scholarship, and perhaps depositing their sources online for other scholars to use. A vigorous public domain today is a prerequisite for a healthy historical record.[65]

More than a century ago, Justin Winsor, the third president of the AHA, concluded his Presidential Address—focused on a topic that would be considered odd today, that of preserving manuscript sources for the study of history—with a plea to the AHA "to convince the National Legislature" to support a scheme "before it is too late" to preserve and make known "what there is still left to us of the historical manuscripts of the country." For founders of the historical profession such as Winsor, the need to engage with history broadly defined—not just how it was researched but also how it was taught in the schools or preserved in archives—came naturally; it was part of creating a historical profession.[66] In the early twenty-first century, we are likely to be faced with re-creating the historical profession, and we will be well served by such a broad vision of our mission. If the past is to have an abundant future, if the story of Bert Is Evil! and hundreds of other stories are to be fully told, then historians need to act in the present.

2

Web of Lies?

Historical Knowledge on the Internet

with Daniel J. Cohen

In the spring of 2004 when the *New York Times* decided to offer an assessment of the social and cultural significance of Google (then heading toward its highly successful IPO), it provided the usual sampling of enthusiasts and skeptics. The enthusiasts gushed about the remarkable and serendipitous discoveries made possible by Google's efficient search of billions of Web pages. Robert McLaughlin described how he tracked down five left-handed guitars robbed from his apartment complex. Orey Steinmann talked about locating the father he had been stolen from (by his mother). And a New York City woman unearthed an outstanding arrest warrant for a man who was courting her.

Perhaps inevitably the dour skeptics hailed from the academic and scholarly world—and particularly from one of its most traditional disciplines, history. Bard College President Leon Botstein told the *Times* that a Google search of the Web "overwhelms you with too much information, much of which is hopelessly unreliable or beside the point. It's like looking for a lost ring in a vacuum bag. What you end up with mostly are bagel crumbs and dirt." He cautioned that finding "it on Google doesn't make it right." Fellow historian and Librarian of Congress James Billington nodded in sober agreement: "far too often, it is a gateway to illiterate chatter, propaganda and blasts of unintelligible material."[1]

Botstein and Billington were just the latest in a long line of historians who have viewed the Internet with substantial skepticism. In November 1996, for example, British historian Gertrude Himmelfarb offered a "neo-Luddite" critique of the then relatively young Web in the *Chronicle of Higher Education*.[2] Five years later, in a *Journal of American History* round table on teaching American history survey courses, several participants expressed similar reservations. "Luddite that I am, I do not use Web sites or other on-line sources, in part because I'm not up on them, but also because I'm old-fashioned enough to believe that there is no substitute for a thick book and an overstuffed chair," Le Moyne College professor Douglas Egerton confessed. He harbored serious concerns about the effect of the new medium on his students' historical literacy: "Many of my sophomores cannot distinguish between a legitimate Web site that has legitimate primary documents or reprinted (refereed) articles and pop history sites or chat rooms where the wildest conspiracies are transformed into reality." Elizabeth Perry, a professor at St. Louis University and non-Luddite, said that she thought that "the Internet can be a wonderful resource" for historical materials. But like Egerton she had major doubts about using it in the classroom: "I find that students do not use [the Web] wisely. They accept a great deal of what they see uncritically . . . [and] when they can't find something on the Web, they often decide that it doesn't exist." Sensing that the Web is filled with inaccurate "pop" history that pulls students away from the rigorous historical truth found in "thick books," only two of the 11 participants in the *Journal of American History* round table admitted to using Web resources regularly in their history surveys, and one of those professors stuck with the textbook publisher's Web site rather than venturing out to the wilds of the broader Web.[3] This sample of professional historians is not atypical; in a recent study, a scant 6 percent of instructors of American history survey courses put links to the Web on their online syllabi (beyond perfunctory links to official textbook Web sites).[4]

Is this skepticism merited? What is the quality and accuracy of historical information on the Web? With Google now indexing more than eight billion pages, a full qualitative assessment of historical information and writing on the Web is well beyond the ability of any person or even team of people. It is, in fact, akin to proposing to assess all the historical works in Billington's own

Library of Congress. Faced with that fool's errand, we take a very different approach to assessing the quality of historical information on the Web, one that relies on two of its most distinctive qualities—its massive scale and the way that its contents can be rapidly scanned and sorted. These are the qualities that are central to Google's extraordinary success as a swift locator of people and information. And, in fact, we employ Google as our indispensable assistant for assessing the veracity of the Web's historical information. We also argue, somewhat counterintuitively, that Botstein may be right about the Web containing a lot of dust and bagel crumbs, while at the same time being wrong in his overall claim about the Web's unreliability.

Equally important, we seek to show that our approach suggests in some relatively primitive ways the possibilities for the automatic discovery of historical knowledge, possibilities that historians have tended to discount after their brief flirtation with quantitative history in the 1970s. We conclude with some speculations on the larger implications for historical research and teaching of these claims about the historical reliability of the Web and these methods for demonstrating its reliability. The rise of the Web and the emergence of automated methods for mining historical knowledge digitally are more important for how they may change teaching and research tomorrow than for the gems they may allow us to find among the bagel crumbs today.

How Computer Scientists Differ from Humanists in Their View of the Web

The enormous scale and linked nature of the Web—an unprecedented development—makes it possible for the Web to be "right" in the aggregate while sometimes very wrong on specific pages. This is actually a pragmatic understanding of the Web that underlies much recent work by computer scientists (including those at Google) who are trying to forge a trustworthy resource for information out of the immense chaos of billions of heterogeneous electronic documents. "The rise of the world-wide-web has enticed millions of users," observe Paul Vitanyi and Rudi Cilibrasi of the Centrum voor Wiskunde en Informatica in the Netherlands, "to type in trillions of characters to create billions of web pages of on average low quality contents." Yet, they continue, "the sheer mass of the information available about almost every conceivable topic makes it likely that extremes will cancel and the majority or average is meaningful in a low-quality approximate sense."[5] In other words, while the Web includes many poorly written passages, often uploaded by unreliable or fringe characters, taken as a whole the medium actually does quite a good job encoding accurate, meaningful data. Critics like Himmelfarb and Billington point to specific trees (Web pages) that seem to be ailing or growing in bizarre directions; here we would like to join computer scientists like Vitanyi and Cilibrasi in emphasizing the overall health of the vast forest (the World Wide Web in general). Moreover, we agree with a second principle of information theory that underlies this work: as the Web grows, it will become (again, taken as a whole) an increasingly accurate transcription of human knowledge.

Perhaps we should not be surprised at this divide between the humanist skeptics and the sanguine information scientists. The former are used to analyzing carefully individual pieces of evidence (like the accuracy of a single folio in an archive) while the latter are used to divining the ties, relationships, and overlaps among huge sets of documents. Computer scientists specialize in areas such as "data mining" (finding statistical trends), "information retrieval" (extracting specific bits of text or data), and "reputational systems" (determining reliable documents or actors), all of which presuppose large corpuses on which to subject algorithms. Despite a few significant, though short-lived, collective affairs with databases and quantitative social science methods, most contemporary scholars in the humanities generally believe that meaning is best derived by an individual reader (or viewer in the case of visual evidence) and expressed in prose rather than the numbers algorithms produce. Computer scientists use digital technologies to find meaningful patterns rapidly and often without human intervention; humanists believe that their disciplines require a human mind to discern and describe such meaning.

Indeed, Vitanyi and Cilibrasi largely make their case about the "meaningful" nature of the Web using some abstruse mathematical analyses. But they also present a few examples that laypeople can understand and might find surprising. For instance, by feeding just the titles of fifteen paintings into Google, Vitanyi and Cilibrasi's computer program was able to sort these works very accurately into groups that corresponded with their different painters—in this case, one of three seventeenth-century Dutch artists: Rembrandt van Rijn, Jan Steen, or Ferdinand Bol. The program, and the search engine it used for its data, Google, obviously did not have eyes for the fine distinctions between the brush styles of these masters; instead, through some swift mathematical calculations about how frequently these titles showed up on the same pages together (common "hits" on Google) the program calculated that certain titles appeared "closer" than others in the universe of the Web. Thus, Vitanyi and Cilibrasi theorized (based on some statistical principles of information that pre-date the Internet) that the same artist painted these "close" titles. Even if some Web pages had titles from more than one painter (a very common occurrence), or, more troublingly (and perhaps also common on the Web), some Web pages erroneously claimed that Rembrandt painted *Venus and Adonis* when in fact it was Bol, the overall average in an enormous corpus of Web pages on seventeenth-century Dutch painting is more than good enough to provide the correct answer.

Does the same hold true for history? Is the average of all historical Web pages "meaningful" and accurate? To answer this question, one of us (Dan) developed an automated historical fact finder called "H-Bot" beginning in spring 2004, which is available in an early release on the Center for History and New Media (CHNM) Web site at http://chnm.gmu.edu/tools/h-bot. (Computer scientists call software agents that scan the Internet "bots"; the "H," of course, is for "history.") H-Bot is a more specialized version of the information retrieval software that database and Internet companies as well as the U.S. government have been feverishly working on to answer questions automatically that would have required enormous manual sifting of documents in the pre-digital era.[6] (Yes, many of these applications have import for intelligence units like the CIA and NSA.) Although by default the publicly available version looks at "trusted sources" such as encyclopedias first (a method now used by Google, MSN, and other search engines in their recent attempts to answer questions reliably and directly rather than referring users to relevant Web pages), we also have a version that relies purely on a statistical analysis of the Google index to answer historical queries. In other words, we have been able to use H-Bot to assess directly whether we can extract accurate historical information from the dust and bagel crumbs of the Web. And understanding how H-Bot works, as well as the strengths and weaknesses of its methodology, provides a number of important insights into the nature of online knowledge.

H-Bot, a History Software Agent

Suppose you were curious to know when the French impressionist Claude Monet moved to Giverny, the small village about forty-five miles west of

Paris where he would eventually paint many of his most important works, including his famous images of water lilies. Asking H-Bot "When did Monet move to Giverny?" (in its "pure" mode where it does not simply try to find a reliable encyclopedia entry on the matter) would prompt the software to query Google for Web pages from its vast index that include the words "Monet," "moved," and "Giverny"—approximately 6,200 pages in April 2005. H-Bot would then scan the highest ranking of these pages—that is, the same ones that an uninformed student is likely to look at when completing an assignment on Monet—as a single mass of raw text about Monet. Breaking these Web pages apart into individual words, it would look in particular for words that look like years (i.e., positive three- and four-digit numbers), and indeed it would find many instances of "1840" and "1926" (Monet's birth and death years, which appear on most biographical pages about the artist). But most of all it would find a statistically indicative spike around "1883." As it turns out, 1883 is precisely the year that Monet moved to Giverny.

Out of the pages it looked at, H-Bot found some winners and some losers. In fact, a few Web pages at the very top of the results on Google—ostensibly the most "reputable" pages according to the search engine's ranking scheme—get the Giverny information wrong, just as the Web's detractors fear. In this case, however, some of the incorrect historical information does not come from shady, anonymous Internet authors posing as reliable art historians. A page on the Web site of the highly respectable Art Institute of Chicago—the twelfth page H-Bot scanned to answer "When did Monet move to Giverny?"—erroneously claims that "Monet moved to Giverny almost twenty years" after he left Argenteuil in 1874.[7] To be fair to the Art Institute, another of their Web pages that ranked higher than this page gets the date right. But so do the vast majority (well over 95 percent) of the pages H-Bot looked at for this query. In the top 30 Google results, the official Web site of the village of Giverny, the French Academie des Beaux-Arts, and a page from the University of North Carolina at Pembroke correctly specify 1883; so does the democratic (and some would say preposterously anarchical) Web reference site *Wikitravel*, the Australian travel portal *The Great Outdoors*, and CenNet's "New Horizons for the Over 50s" lifestyle and dating site.[8] Through a combination of this motley collection of sites (some might say comically so), H-Bot accurately answers the user's historical question. The Web's correct historical information overwhelms its infirmities. Moreover, the combinatorial method of the software enables it to neutralize the problems that arise in a regular Google search that focuses on the first few entries in the hope that a randomly selected highly ranked link will provide the correct answer (otherwise known as the lazy student method).

Right now H-Bot can only answer questions for which the responses are dates or simple definitions of the sort you would find in the glossary of a history textbook. For example, H-Bot is fairly good at responding to queries such as "What was the gold standard?", "Who was Lao-Tse?", "When did Charles Lindbergh fly to Paris?", and "When was Nelson Mandela born?" The software can also answer, with a lower degree of success, more difficult "who" questions such as "Who discovered nitrogen?" It cannot currently answer questions that begin with "how" or "where," or (unsurprisingly) the most interpretive of all historical queries, "why." In the future, however, H-Bot should be able to answer more difficult types of questions as well as address the more complicated problem of disambiguation—that is, telling apart a question about Charles V the Holy Roman Emperor (1500–1558) from one about Charles V the French king (1338–1380). To be sure, H-Bot is a work in progress, a young student eager to learn. But given that its main programming has been done without an extensive commitment of time or resources by a history professor and a (very talented) high-school student, Simon Kornblith, rather than a team of engineers at Google or MIT, and given that a greater investment would undoubtedly increase H-Bot's accuracy, one suspects that the software's underlying principles are indicative of the promise of the Web as a storehouse of information.

Looking not merely at the answers H-Bot provides but at its "deliberations" (if we can anthropomorphically call them that) provides further insights into the nature of historical knowledge of the Web. Given its enormity compared to even a large printed reference work like the *Encyclopedia Britannica*, the writing about the past on the Web essentially functions as a giant thesaurus, with a wide variety of slightly different ways of saying the same thing about a historical event. For instance, H-Bot currently doesn't get "When did War Admiral lose to Seabiscuit?" right, but it does get "When did Seabiscuit defeat War Admiral?" correct. It turns out that Web authors have thus far chosen the latter construction—"Seabiscuit defeated War Admiral"— but not the former construction to discuss the famous horseracing event of 1938. Of course, given the exponential growth of the Web and the recent feverish interest in Seabiscuit, H-Bot will likely get "When did War Admiral lose to Seabiscuit?" right in the near future, perhaps as soon as a high school student doing a report on the Great Depression writes the phrase "War Admiral lost to Seabiscuit in 1938" and posts it to his class's modest Web site.

Testing H-Bot and the Web

H-Bot is still an unfunded research project in beta release rather than a fully developed tool. Yet even in its infancy, it is still remarkably good at answering historical questions, and its accuracy is directly tied to the accuracy of historical information on the Web. Furthermore, as Vitanyi and Cilibrasi speculate based on the mathematics of information theory, as the Web grows it will become increasingly accurate (as a whole) about more and more topics. But how good is it right now in comparison to an edited historical reference work?

We first conducted a basic test of H-Bot against the information in *The Reader's Companion to American History*, a well-respected encyclopedic guide edited by the prominent historians Eric Foner and John A. Garraty.[9] Taking the first and second biographies for each letter of the alphabet (except "X," for which there were no names), we asked H-Bot for the birth year of the first figure and the death year of the second figure—perhaps some of the most simple and straightforward questions one can ask of the software. For 48 of these 50 questions H-Bot gave the same answer as *The Reader's Companion*, for an extremely competent score of 96 percent. And a closer look at the two divergent results indicates that H-Bot is actually even more accurate than that.

H-Bot's first problem came in answering the question "When did David Walker die?" Rather than offer the answer (1830) provided by *The Reader's Companion*, it replies politely "I'm sorry. I cannot provide any information on that. Please check your spelling or rephrase your query and try again." But no rewording helps since H-Bot has no way of distinguishing the abolitionist David Walker (the one listed in *The Reader's Companion*) from various other "David Walkers" who show up in the top-ranking hits in a Google search—the Web designer, the astronaut, the computer scientist at Cardiff University, the Web development librarian, and the Princeton computer scientist. This disambiguation problem is one of the thorniest issues of information retrieval and data mining in computer science. Perhaps because he is familiar with the problem, the Cardiff David Walker even offers his own disambiguation Web page in which he explains, "There are many people called David

Walker in the world" and notes that he is not the Princeton David Walker nor yet a third computer scientist named David Walker (this one at Oxford).[10] Despite the importance of the disambiguation problem, it is not a problem of the quality of historical information on the Web but rather of the sophistication of the tools—like H-Bot—for mining that information.

The other answer H-Bot gave that differed from *The Reader's Companion* is more revealing in assessing the quality of historical information online. The first name in "H" was Alexander Hamilton, so we asked the software "When was Alexander Hamilton born?" It answered 1757, two years later than the date given in *The Reader's Companion*. But behind the scenes H-Bot wrung its hands over that other year, 1755. Statistically put, the software saw a number of possible years on Web pages about Hamilton, but on pages that discuss his birth there were two particularly tall spikes around the numbers 1755 and 1757, with the latter being slightly taller among the highest-ranked pages on the Web (and thus H-Bot's given answer).

Although one might think that the birth year of one of America's Founders would be a simple and unchanging fact, recent historical research has challenged the commonly mentioned year of 1755 used in the 1991 *Reader's Companion* by the well-known historian Edward Countryman, the author of the Hamilton profile. Indeed, writing just under a decade later in *American National Biography*, the equally well-known historian Forrest McDonald explains, "the year of birth is often given as 1755, but the evidence more strongly supports 1757." A recent exhibit at the New-York Historical Society (N-YHS) decided on the later year as well, as have many historians—and also thousands of Web articles on Hamilton.[11] Indeed, newer Web pages being posted online more often use 1757 than 1755 (including the Web site for N-YHS's exhibit), providing H-Bot with an up-to-the-minute "feeling" of the historical consensus on the matter. This sense based on statistics also means that H-Bot is more up-to-date than *The Reader's Companion*, and with the Web's constant updating the software will always be so compared to any such printed work. (The same constant updating as well as continual changes in Google's index means that some of the queries we discuss here may produce different results at different times.)

Another, perhaps more accurate way of understanding H-Bot is to say that the software is concerned not with what most people would call "facts" but rather with consensus. The Web functions for its software as a vast chamber of discussion about the past. This allows H-Bot to be in one way less authoritative but in another way—as in the case of Hamilton's birth year—more flexible and current. Such "floating" facts are more common in history (and indeed the humanities in general) than nonprofessionals might suspect, especially as one goes further back into the dark recesses of the past. Professional historians have revised the commonly accepted birth year of Genghis Khan several times in the past century; even more recent subjects, such as Louis Armstrong and Billie Holiday, have "facts" such as their birth years under dispute.

The same algorithms, however, that allow H-Bot to answer swiftly and accurately questions such as birth and death years reveal a weakness in the software. Take, for instance, what happens if you ask H-Bot when aliens landed in Roswell or when Stalin was poisoned, two common rumors in history. H-Bot "correctly" answers these questions as 1947 and 1953, respectively. It statistically analyzes the many Web pages that discuss these topics (a remarkable quarter-million pages in the case of the Roswell aliens) to get the agreed-upon years it returns as answers. In these cases, the "wisdom of crowds" turns into the "madness of crowds."[12]

Conceivably this infirmity in H-Bot's algorithms—we might call it historical gullibility or naiveté—could be remedied with more programming resources. For instance, it may be that when a topic is discussed on many Web pages that end in .com but not on many pages that end in .edu (compared to the relative frequency of those top-level domains on the Web in general), the program could raise a flag of suspicion. Or H-Bot could quickly analyze the grade level of the writing on each Web page it scanned and factor that into its "reliability" mathematics. Both of these methods have the potential to reinject a measure of antidemocratic elitism that the Web's critics see as missing in the new medium. Further research into this question of rumor versus fact would benefit not only H-Bot but also our understanding of the popular expression of history on the Web.

More broadly, however, the problem reflects the reality that even the "facts" in history can be a matter of contention. After all, lots of people genuinely and passionately believe that aliens did land at Roswell in 1947. And a very recent scholarly book argues that Stalin was poisoned in 1953.[13] If that becomes the accepted view, the Web is well positioned to pick up the shifting consensus more quickly than established sources. This also reminds us of the ways that the Web challenges our notions of historical "consensus"—broadening debates from the credentialed precincts of the American Historical Association's Annual Meeting (and its publications) to the rough-and-tumble popular debates that occur online.

Following its duel with *The Reader's Companion to American History*, we then fed into H-Bot's software a larger and more varied list of historical questions, though still with the goal of seeing if it could correctly answer the year in which the named event occurred. For this test we used an edition of *The Timetables of History* (1982), a popular book by Bernard Grun that features a foreword by Daniel J. Boorstin, the former Librarian of Congress. The first event listed in every third year from 1670 to 1970 was rephrased as a question and then fed into H-Bot. The questions began with the now somewhat obscure Treaty of Dover (1670) and the Holy Roman Emperor Leopold I's declaration of war on France (1673) and ended with the dates for the Six-Day War (1967) and the shooting of students by the National Guard at Kent State University (1970). Although skewed more toward American and European history, many of the 100 questions covered other parts of the globe.

H-Bot performed respectably on this test but did less well than on the test of major American figures, correctly providing the year listed in the reference book for 74 of the 100 events. The software could not come up with an answer for three of the 100 questions and was wildly wrong on another five. But it came within one year of the *Timetables* entry for another eight answers, within two years on one entry, and was within 50 years for another nine (mostly, in those cases, choosing a birth or death year rather than the year of an event that a historical actor participated in). A more generous assessment might therefore give H-Bot 83 percent, a respectable B or B-, on this test, if one included the nine near misses but excluded the nine distant misses.

H-Bot's difficulties with these nine distant misses and the five completely wrong answers were again due, in this case almost entirely, to disambiguation issues—for example, the ability to understand when a question about Frederick III refers to the Holy Roman Emperor (1415–1493), the Elector of Saxony (1463–1525), or the King of Denmark and Norway (1609–1670). Many of the questions we gave to H-Bot unsurprisingly dealt with royalty whose names were far from distinctive, especially in European history. Through an added feedback mechanism—responding to a query with "Do you mean Frederick III the Holy Roman Emperor, the Elector of Saxony, or the King of Denmark and Norway?"—H-Bot could tackle the problem of disambiguation, which seems to have flummoxed the software in those eight cases ranging from Leopold I's declaration of war on France in 1673 to Tewfik Pasha's death in 1892. (The expansive, populist online encyclopedia *Wikipedia* helpfully has "disambiguation pages," often beginning with a variant of "There are more than one of these . . . " to direct confused researchers to the entry they are looking for; the reference site Answers.com uses these pages to send researchers in the correct direction rather than rashly providing an incorrect answer as H-Bot currently does.) A second, lesser problem stems from the scarcity of Web pages on some historical topics, and as we noted earlier, this problem may solve itself. As the Web continues its exponential growth and the topics written on it by professional and amateur historians proliferate, the software will find more relevant pages on, say, late seventeenth-century politics than it is currently able to do.

Even with these caveats, one might in common sense turn off the computer and head for the reference bookshelf to find hard facts if the best H-Bot can do is provide three precisely correct answers for every four questions. Yet this test of H-Bot also revealed significant infirmities with reference works like *The Timetables of History*—infirmities that are less obvious to the reader than to the online researcher using H-Bot, which is able to instantly "show its work" by identifying the relevant Web pages it looked at to answer a query. Most people consider reference books perfect, but we should remember what our fourth-grade teachers told us: Don't believe everything you read. *Timetables* suffers from disambiguation problems as well, albeit in a less noticeable way than H-Bot. For instance, in its 1730 entry, the book says that "Ashraf" was killed. But which Ashraf? *Wikipedia* lists no fewer than twenty-five Ashrafs, mostly Egyptian sultans but also more recent figures such as Yasser Arafat's doctor. This 1730 entry, like many others in *Timetables*, doesn't really help the ignorant user.

More seriously, although currently better than H-Bot, *Timetables* incorrectly identified at least four "facts" wrong in the test, albeit in minor ways. The reference work claims that the state of Delaware separated from Pennsylvania in 1703; it actually gained its own independent legislature in 1704. Great Britain acquired Hong Kong as a colony in 1842 (the Treaty of Nanking following the First Opium War), not 1841. Kara Mustafa, the Ottoman military leader, died in 1683, not 1691 as the book claims. *Timetables* also claims that Frederick III of Prussia was crowned as king of that German state in 1688, but he was really crowned as Elector of Brandenburg in that year, only later becoming king of Prussia in 1701—the year correctly given by H-Bot. These are lesser errors, perhaps, than the gross errors H-Bot sometimes makes. Still, it reminds us that getting all the facts right is more difficult than it looks. "People don't realize how hard it is to nail the simplest things," Lars Mahinske, a senior researcher for the authoritative *Encyclopedia Britannica* confessed to *Wall Street Journal* reporter Michael McCarthy in 1990. But, unlike *Britannica* (or at least its traditional print form), H-Bot, as we have noted, can be self-correcting whereas a print book is forever fixed, errors and all.

Factualist History: H-Bot Takes the National Assessment History Exam and Acquires Some New Talents

Some might object that H-Bot is not a true "historian" and that it can only answer narrowly factual questions that serious historians regard as trivial. But before we dismiss H-Bot as an idiot savant, we should observe that these are precisely the kinds of questions that policy makers and educators use as benchmarks for assessing the state of student (and public) "ignorance" about the past—the kinds of questions that fill "Standards of Learning" (SOL) and National Assessment of Educational Progress (NAEP) tests.

We decided to have a special multiple-choice test-taking version of H-Bot (currently unavailable to the public) take the NAEP United States History examination to see if it could pass. (The NAEP exam actually consists of both multiple-choice questions and short answer questions; since H-Bot has not yet become as responsive and loquacious as *2001: A Space Odyssey*'s HAL 9000, we only had it take the multiple-choice questions.) This test-taking version of the software works by many of the same principles as the historical question-answering version, but it adds another key technique ("normalized information distance") from the computer scientists' toolkit that, as will be explained below, opens up some further possibilities for the automatic extraction of historical information. It is the use of "normalized information distance," in fact, that enables Vitanyi and Cilibrasi to automatically associate seventeenth-century Dutch paintings with their painters without actually viewing the works.

The test-taking H-Bot gains a significant advantage that human beings also have when they take a multiple-choice exam—and which makes these sorts of tests less than ideal assessments of real historical knowledge. In short, unlike the open-ended question "When did Monet move to Giverny?", multiple-choice questions obviously specify just three, four, or five possible responses. By restricting the answers to these possibilities, multiple-choice questions provide a circumscribed realm of information where subtle clues in both the question and the few answers allow shrewd test-takers to make helpful associations and rule out certain answers (test preparation companies like Stanley Kaplan and Princeton Review have known this for decades and use it to full advantage). This "gaming" of a question can occur even when the test-taker doesn't know the correct answer and is not entirely familiar with the historical subject matter involved. But even with this inherent flaw, the questions on the NAEP can be less than straightforward and reflect a much harder assignment for H-Bot, and one that challenges even the enormous database of historical writing on the Web. To be sure, some of the questions are of a simplistic type that the regular H-Bot excels at: a one-word answer, for instance, to a straightforward historical question. For example, one sample question from the NAEP fourth-grade examination in American history asks how "Most people in the southern colonies made their living," with the answer obviously being "farming" (rather than "fishing," "shipbuilding," or "iron mining"). H-Bot quickly derives this answer from the plethora of Web pages about the antebellum South by noting that this word appears far more often than the other choices on pages that mention "Southern" and "colonies."

More generally, and especially in the case of more complicated questions and multiword answers, the H-Bot exam-taking software tries to figure out how closely related the significant words in the question are to the significant words in each possible answer. H-Bot first breaks down the question and possible responses into their most significant words, where significance is calculated (fairly crudely, but effectively, in a manner similar to Amazon. com's "statistically improbable phrases" for books) based on how infrequently a word appears on the Web. Given the question "Who helped to start the boycott of the Montgomery bus system by refusing to give up her seat on a segregated bus?" H-Bot quickly determines that the word "bus" is fairly common on the Web (appearing about 73 million times in Google's index) while the word "montgomery" is less so (about 20 million instances of the many different meanings of the word). Similarly, in possible answer (c), "Rosa Parks," "parks" is found many places on the Web while "rosa" appears on only a third as many pages. Having two uncommon words such as "montgomery" and "rosa" show up on a number of Web pages together seems even more unusual—and thus of some significance. Indeed, that is how H-Bot figures out that the woman who refused to go to the back of the bus, sparking the Montgomery, Alabama boycott and a new phase of the civil rights movement, was none other than Rosa Parks (rather than one of the three other possible answers, "Phyllis Wheatley," "Mary McLeod Bethune," or "Shirley Chisholm").

As computer scientists Vitanyi and Cilibrasi would say, this high coincidence of "rosa" and "montgomery" in the gigantic corpus of the Web means that they have a small "normalized information distance," an algorithmic measure of closeness of meaning (or perhaps more accurately, a measure of the lack of randomness in these words' coincidence on the Web).[14] In more humanistic terms, assuming that Google's index of billions of Web pages encodes relatively well the totality of human knowledge, this particular set of significant words in the right answer seems more closely related to those in the question than the significant words in the other, incorrect answers.

H-Bot can actually assess the normalized information distance between sets of words, not just individual words, in the question and possible answers, increasing its ability to guess correctly using Vitanyi and Cilibrasi's "web pages of on average low quality contents." These algorithms combined with the massive corpus of the Web allow the software to swiftly answer questions on the NAEP that supposedly invoke the higher-order processes of historical thinking and that should be answerable only if you truly understand the subject matter and are able to reason about the past. For example, a NAEP question asks "What is the main reason the Pilgrims and Puritans came to America?" and provides the following options:

(a)   To practice their religion freely
(b)   To make more money and live a better life
(c)   To build a democratic government
(d)   To expand the lands controlled by the king of England

H-Bot cannot understand the principles of religious freedom, personal striving, political systems, or imperialism. But it need not comprehend these concepts to respond correctly. Instead, to answer the seemingly abstract question about the Puritans and Pilgrims and why they came to America, H-Bot found that Web pages on which words like "Puritan" and "Pilgrim" appear contain the words "religion" and (religious) "practice" more often than words like "money," "democratic," or "expand" and "lands." (To be more precise, in this case H-Bot's algorithms actually compare the normal frequency of these words on the Web with the frequency of these words on relevant pages, therefore discounting the appearance of "money" on many pages with "Puritan" and "Pilgrim" because "money" appears on over 280 million Web pages, or nearly one out of every 28 Web pages.) H-Bot thus correctly surmises that the answer is (a). Again, using the mathematics of normalized information distance the software need not find pages that specifically discuss the seventeenth-century exodus from England or that contain an obvious sentence such as

"The Puritans came to America to practice their religion more freely." Using its algorithms on various sets of words, it can divine that certain combinations of rare words are more likely than others. It senses that both religion and freedom had a lot to do with the history of the Pilgrims and Puritans.

Or perhaps we should again be a little more careful (or some might say more cynical) and say that this assessment of the centrality of religious liberty is the reigning historical interpretation of the Pilgrims and Puritans, rather than a hard and uncontroversial fact supported by thousands of Web pages. While most amateur and professional historians would certainly agree with this assessment, we could find others who would disagree, advancing economic, political, or imperialistic rationales for the legitimacy of answers (b), (c), or (d). Yet these voices are overwhelmed on the Web by those who hew closely to the textbook account of the seventeenth-century British emigration to the American colonies. Undoubtedly, conservatives would be pleased with this online triumph of consensus over interpreters of the past who dare to use Marxist lenses to envision the founding of the United States.[15] But challenging conventional and textbook accounts often forms an important part of understanding the past more fully. For instance, Charles Mann's recent book on pre-Columbian America shows that virtually all of the information on the age, sophistication, and extent of American Indian culture in U.S. history survey textbooks is out of date.[16] Thus, the factualist H-Bot offers an impoverished view of the past—just like high school textbooks and standardized tests.

Interestingly, the use of normalized information distance to identify strong consensus also allows H-Bot to answer uncannily some questions (which we did not include in the main testing) that seem to require visual interpretation. For example, although it cannot see the famous picture of Neil Armstrong next to an American flag on the surface of the moon, H-Bot had no trouble correctly answering the question below this photograph:

What is the astronaut in this picture exploring?
(a)   The Sun
(b)   The Arctic
(c)   The Moon
(d)   Pluto

There are far fewer pages on which the word "astronaut" appears along with either "arctic" or "Pluto" than it does with "moon." "Astronaut" shows up on roughly the same number of Web pages with "sun" as it does with "moon" (11,340 pages versus 11,876), but since overall the number of Web pages that mention "moon" is about one-quarter the number of pages that mention "sun" (about 20 million versus 75 million), H-Bot understands that there is a special relationship between the moon and astronauts. Put another way, when a Web page contains the word "astronaut," the word "moon" is far more likely to appear on the page than its normal frequency on the Web. H-Bot therefore doesn't need to "understand" the Armstrong photograph or its history *per se*, or know that this event occurred in 1969—the documents in the massive corpus of the Web point to the overwhelming statistical proximity of answer (c) to the significant words in the question.

As with its other incarnation, H-Bot shows imperfections in taking the NAEP test. These infirmities differ from the problems related to rumors and falsehoods in the main edition of the software. Since, as the moon question shows, H-Bot is hard-wired to look for close rather than distant relationships, it has trouble with the occasional NAEP question that is phrased negatively, i.e., "which one of these is NOT true. . . ?" Also, when it is asked which event comes first or last in a chronological series, the algorithm blindly finds the most commonly discussed event. Again, such problems could be largely remedied with further programming, and they reflect the imperfections of H-Bot as a tool rather than the imperfections of online historical information.

Moreover, even with these weaknesses, when H-Bot was done processing the 33 multiple-choice questions from the fourth-grade NAEP American history exam, it had gotten 27 answers right, a respectable 82 percent. And as with the other version of H-Bot, when a larger (and more updated) index of the Web is available, it should do even better on these exams. The average fourth grader does far worse. In 1994, the most recent year for this test data, 69 percent knew of the American South's agrarian origins, 62 percent identified Rosa Parks correctly, and a mere 41 percent understood the motivation behind the emigration of the Puritans and Pilgrims.

Conclusion: Automatic Discovery and the Study of History

Scientists have grown increasingly pessimistic about their ability to construct the kind of "artificial intelligence" that was once widely seen as lurking just around the next computer.[17] In every decade since the Second World War, technologists have predicted the imminent arrival of a silicon brain equivalent to or better than a human's, as the remarkable pace of software innovation and hardware performance accelerates. Most recently, the technology magnate Ray Kurzweil asserted in *The Age of Spiritual Machines* (2000) that by 2029 computer intelligence will not only have achieved parity with the human mind; it will have surpassed it. Since the first attempts at machine-based reasoning by the Victorians, however, the mind repeatedly has proven itself far more elusive than technologists have expected. Kurzweil is surely correct that computers are becoming more and more powerful each year; yet we remain frustratingly distant from a complete understanding of what we're trying to copy from the organic matter of the brain to the increasingly speedy pathways of silicon circuits.

But while the dawn of fully intelligent machines seems to constantly recede into the distance, the short-term future for more circumscribed tools like H-Bot looks very bright. One reason is that the Web is getting very big, very fast. In September 2002, Google claimed to have indexed 2 billion Web pages. Just 15 months later, it was boasting about 4 billion. And it only took 11 more months to double again to 8 billion.[18] These numbers matter a great deal because, as we have noted, the power of automatic discovery rests as much on quantity as quality. (Indeed, H-Bot works off of Google's public application programming interface, or API, which permits access to only about 1.5 billion pages in its index; the company seems wary of providing access to the full 8 billion.)

But, as it happens, the quality of what's out on the public Web is also going to improve very shortly because Google is busy digitizing millions of those "thick books" that we generally read in "overstuffed chairs." If automatic discovery can do so well working from the Web pages of high schoolers, think what it can do when it can prowl through the entire University of Michigan library. Finally, the algorithms for extracting meaning and patterns out of those billions of pages are getting better and better. Here, too, Google (and its competitors at Yahoo and Microsoft) are responsible. After all, a humble history Ph.D., trained in the history of science rather than the science of algorithms, created H-Bot. But Google and its competitors are hiring squadrons of the most talented engineers and Ph.D.s, from a variety of fields such as computer science, physics, and of course mathematics, to work on automatic discovery methods. In the second quarter of 2005 alone, Google added 230 new programmers to improve its vast search empire; one computer science professor noted that the top third of his students in a class on search technology went straight to Google's awaiting offers. Yahoo is not far behind in this arms race, and Microsoft is devoting enormous resources to catching up.[19] We are at a unique moment in human history when literally millions of dollars are being thrown into the quest for efficient mathematical formulas and techniques to mine the billions of documents on the Internet.

So H-Bot may only be able to score 82 percent on the NAEP U.S. History examination today; its successors are likely to reach 95 or even 99 percent. Should we care? Most historians already know about the Pilgrims or at least know how to find out quickly. Why should a software agent that can answer such questions impress them? Isn't this simply a clever parlor trick? We think that the answer is no, because these developments have significant implications for us as teachers and researchers.

H-Bot may never become "intelligent," but it has already proven itself very smart at answering multiple-choice history questions. Most historians would likely view H-Bot's factual abilities with disdain or condescension. Yet, paradoxically, we (and our compatriots in the world of education) have often decided that precisely such abilities are a good measure of historical "knowledge." As history educator and psychologist Sam Wineburg has shown, we have spent the past 90 years bemoaning the historical "ignorance" of students based on their "[in]ability to answer factual questions about historical personalities and events." And the conclusions that have been drawn from that inability have been nothing short of apocalyptic. In 1943, the *New York Times* took time out from its wartime coverage to rebuke "appallingly ignorant" college freshmen for their scores on the national history exam it had administered the year before. Almost a half century later, Diane Ravitch and Chester Finn declared that low scores on the NAEP placed students "at risk of being gravely handicapped by . . . ignorance upon entry into adulthood, citizenship, and parenthood."[20]

As Wineburg points out, it is, in part, technology that is responsible for our attachment to the factual, multiple-choice test. "We use these tests," he writes, "not because they are historically sound or because they predict future engagement with historical study, but because they can be read by machines that produce easy-to-read graphs and bar charts."[21] Any instructor weighed down by a stack of unmarked history essays who sees a colleague walk away smiling from the Scantron machine with his or her grading completed knows this point all too well. But technology may also bring their demise or transformation. A year or two ago, our colleague and fellow historian Peter Stearns proposed the idea of a history analog to the math calculator, a handheld device that would provide students with names and dates to use on ex-

ams—a Cliolator, he called it, a play on the muse of history and the calculator. He observed that there would likely be resistance to the adoption of the
Cliolator, as there had been by some educators to the calculator. But he also
argued, rightly in our view, that it would improve history education by displacing the fetishizing of factual memorization.

When we discussed this idea with Peter, we at first focused on the ways
that the Cliolator would be much harder to build than the calculator. After
all, the calculator only needs to understand some basic principles of mathematics to answer a nearly infinite number of questions. The Cliolator would
need to be loaded with very substantial quantities of historical information
before it would be even modestly useful. Could you possibly anticipate every
conceivable fact a student might need to know? But then we realized that we
were already building a version of the Cliolator in H-Bot and that millions of
people were uploading those millions of historical facts for us onto their Web
pages.

The combination of the magnificent, if imperfect, collective creation of
the Web with some relatively simple mathematical formulas has given us a
free version of the Cliolator. And the handheld device for accessing it—the
cell phone—is already in the pockets of most school kids. In a very short
time, when cell phone access to Web-based materials becomes second nature
and H-Bot (or its successors) gets a little bit better, students will start asking
us why we are testing them on their ability to respond to questions that their
cell phones can answer in seconds. It will seem as odd to them that they can't
use H-Bot to answer a question about the Pilgrims as it is today to a student
who might be told that they can't use a calculator to do the routine arithme-
tic in an algebra equation. The genie will be out of the bottle, and we will
have to start thinking of more meaningful ways to assess historical knowledge or "ignorance." And that goes for not just high school instructors and
state education officials but also the very substantial numbers of college
teachers who rely on multiple-choice questions and Scantron forms.

If the future of historical teaching looks different in the age of dumb (but
fast) machines, what about historical research? For historical researchers, the
question is not whether mathematical techniques help us ferret out historical
facts from vast bodies of historical accounts. Professional historians have
long known how to locate quickly (and more importantly, assess the quality
of) the facts they need. Rather, the issue is whether these same approaches
help to find things in historical *sources*. Can we mine digitized primary sources
for new insights into the past? Here, a key weakness of H-Bot—its reliance

on consensus views—turns out to be a virtue. After all, a key goal of the cultural historian is to ferret out the consensus—or mentalité—of past generations. When we do history, we are generally more interested in finding out what people believed than whether what they believed was "true."

The potential for finding this out through automated methods has become much greater because of the vast quantities of those digitized primary sources that have suddenly become available in the past decade. Even a very partial list is astonishing in its breadth and depth: the Library of Congress's American Memory project presents more than 8 million historical documents. The Making of America site, organized by the University of Michigan and Cornell University, provides more than 1.5 million pages of texts from mid-nineteenth-century books and articles. ProQuest's Historical Newspapers offers the full text of eight major newspapers including full runs of the *New York Times* and the *Los Angeles Times*. Google's digitization effort will dwarf these already massive projects that put a startlingly large proportion of our cultural heritage into digital form. And Google's project has apparently sparked a digitization arms race. Yahoo has announced an "Open Content Alliance" that will, in partnership with libraries and the Internet Archive, digitize and make available public domain books.[22] The European Commission, worried that the EU will be left behind as these two mammoth American companies convert the analog past into the digital future, recently unveiled its own expansive effort to turn the paper in European libraries into electronic bits.

Our ability to extract historical nuggets from these digital gold mines is limited because many of these collections (unlike the public Web) can only be entered through a turnstile with a hefty tariff. In addition, many of them have limited search capabilities. And the terms of access for the Google project remain murky. There is no Google that can search the collections of the "deep" and "private" Web. Even so, historians have already discovered enormous riches through the simplest of tools—word search.

Yet such efforts have not truly taken advantage of the potent tools available to the digital researcher. Many of the methods of digital humanities research are merely faster versions of what scholars have been doing for decades, if not longer. German philologists were counting instances of specific words in the Hebrew and Christian Bibles, albeit in a far more painstaking manner, before the advent of computerized word counts of digital versions of these ancient texts. And long before computer graphics and maps, the Victorian doctor John Snow was able to divine the origins of a London cholera outbreak by drawing marks signifying the addresses of infected people until he saw that they tended to cluster around a water pump.

Could we learn more using more sophisticated automatic discovery techniques such as the statistical tests used by H-Bot or the normalized information distance employed by its test-taking cousin? Historians—perhaps with the exception of their temporary attraction to quantitative history in the 1970s—have tended to view any automated or statistical aids to studying the past with suspicion. They prefer to view their discipline as an art or a craft rather than a science and they believe that the best historians look at "everything" and then reflect on what they have read rather than, for example, systematically sampling sources in the manner of a sociologist. They lionize the heroic labors of those who spent twenty years working through all of the papers of a Great Man. But the combination of the digitization of enormous swaths of the past and the overwhelming documentation available for some aspects of twentieth-century history will likely force them to reconsider whether historians, like sociologists, need to include "sampling"—especially systematic sampling—as part of their routine toolkit.[23]

The digital era seems likely to confront historians—who were more apt in the past to worry about the scarcity of surviving evidence from the past— with a new "problem" of abundance. A much deeper and denser historical record, especially one in digital form, presents us with an incredible opportunity. But its overwhelming size means that we will have to spend a lot of time figuring out how to take advantage of this opportunity—and that we will probably need sophisticated statistical and data mining tools to do so.[24]

These more systematic approaches to mining historical documentation will also need to take advantage of some of the mathematical approaches that we outline here. Eighteenth-century historians, for example, will surely want to count the number of references to "God" and "Jesus" in the writings of Enlightenment thinkers and Founding Fathers (and Mothers). But why not go beyond that to consider the proximity of different terms? Are some writers more likely to use religious language when talking about death than other subjects? Has that changed over time? Are women writers more likely to use words like "love" or "passion" when talking about marriage in the nineteenth century than in the eighteenth century? Such questions would be relatively easy to answer with the mathematical formulas described here—assuming, of course, proper access to the appropriate bodies of digitized sources.

It would be an illusion to see such approaches as providing us with a well-marked path to historical "truth" or "certainty," as was sometimes promised by some of the most enthusiastic promoters of quantitative history in the 1960s and 1970s. History will never be a science, but that doesn't mean that a more systematic use of evidence and more systematic techniques for mining large bodies of evidence would not assist us in our imperfect quest to interpret the past.[25] Arthur Schlesinger Jr. may be right to argue skeptically that "almost all important questions are important precisely because they are not susceptible to quantitative answers."[26] But that doesn't mean that quantitative and systematic methods can't help us to develop those qualitative answers. Historical data mining might be best thought of as a method of "prospecting," of trying out interesting questions and looking for rich veins of historical evidence that we should examine more closely. Indeed, that kind of prospecting is precisely what John Snow was doing when he marked down cholera deaths on a London map.

Finally, although H-Bot currently answers only simple historical questions, the software ultimately suggests the possibility of some considerably more complex ways that one might analyze the past using digital tools. Might it be possible to use other theories from computer science to raise and answer new historical questions? H-Bot uses just a few of the many principles of information theory—normalized information distance, measures of statistical significance, and methods of automated text retrieval.[27] But these are merely the tip of the iceberg. Are there other, perhaps even more revealing theories that could be applied to historical research on a digital corpus?

3

Wikipedia: Can History Be Open Source?

HISTORY IS A DEEPLY INDIVIDUALISTIC CRAFT. THE singly authored work is the standard for the profession; only about 6 percent of the more than 32,000 scholarly works indexed since 2000 in the *Journal of American History*'s comprehensive bibliographic guide, "Recent Scholarship," have more than one author. Works with several authors—common in the sciences—are even harder to find. Fewer than 500 (less than 2 percent) have three or more authors.[1]

Historical scholarship is also characterized by *possessive* individualism. Good professional practice (and avoiding charges of plagiarism) requires us to attribute ideas and words to specific historians—we are taught to speak of "Richard Hofstadter's status anxiety interpretation of Progressivism."[2] And if we use more than a limited number of words from Hofstadter, we need to send a check to his estate. To mingle Hofstadter's prose with your own and publish it would violate both copyright and professional norms.

A historical work without owners and with multiple, anonymous authors is thus almost unimaginable in our professional culture. Yet, quite remarkably, that describes the online encyclopedia known as *Wikipedia*, which contains 3 million articles (1 million of them in English). History is probably the category encompassing the largest number of articles. *Wikipedia* is entirely free. And that freedom includes not just the ability of anyone to read it (a freedom denied by the scholarly journals in, say, JSTOR, which requires an expensive institutional subscription) but also—more remarkably—their freedom to use it. You can take *Wikipedia*'s entry on Franklin D. Roosevelt and put it on your own Web site, you can hand out copies to your students, and you can publish it in a book—all with only one restriction: you may not impose any more restrictions on subsequent readers and users than have been imposed on you. And it has no authors in any conventional sense. Tens of thousands of people—who have not gotten even the glory of affixing their names to it—have written it collaboratively. The Roosevelt entry, for example, emerged over four years as 500 authors made about 1,000 edits. This extraordinary freedom and cooperation make *Wikipedia* the most important application of the principles of the free and open-source software movement to the world of cultural, rather than software, production.[3]

Despite, or perhaps because of, this open-source mode of production and distribution, *Wikipedia* has become astonishingly widely read and cited. More than a million people a day visit the *Wikipedia* site. The Alexa traffic rankings put it at number 18, well above the *New York Times* (50), the Library of Congress (1,175), and the venerable *Encyclopedia Britannica* (2,952). In a few short years, it has become perhaps the largest work of online historical writing, the most widely read work of digital history, and the most important free historical resource on the World Wide Web. It has received gushing praise ("one of the most fascinating developments of the Digital Age"; an "incredible example of open-source intellectual collaboration") as well as sharp criticism (a "faith-based encyclopedia" and "a joke at best"). And it is almost entirely a volunteer effort; as of September 2005, it had two full-time employees. It is surely a phenomenon to which professional historians should attend.[4]

To that end, this essay seeks to answer some basic questions about history on *Wikipedia*. How did it develop? How does it work? How good is the historical writing? What are the potential implications for our practice as scholars, teachers, and purveyors of the past to the general public?

Writing about *Wikipedia* is maddeningly difficult. Because it is subject to constant change, much that I write about *Wikipedia* could be untrue by the time you read this. An additional difficulty stems from its vast scale. I cannot claim to have read the 500 million words in the entire *Wikipedia*, nor even the subset of articles (as many as half) that could be considered historical.[5] This is only a very partial and preliminary report from an ever-changing front, but one that I argue has profound implications for our practice as historians.

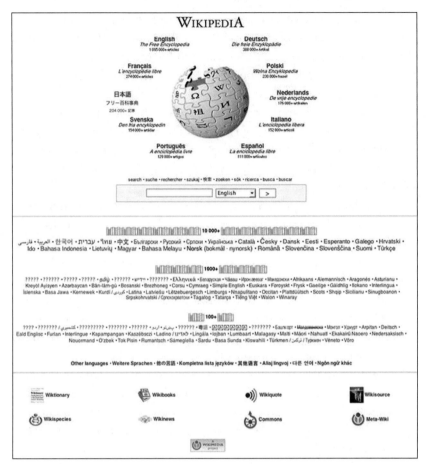

FIGURE 3.1 *Wikipedia* today: This home page for *Wikipedia* reflects the scale of the project (more than 1 million English-language articles) and its multiple languages.
http://wikipedia.org/ *(March 8, 2006)*

Origins

Wikipedia itself rather grandly traces its roots back to "the ancient Library of Alexandria and Pergamon" and the "concept of gathering all of the world's knowledge in a single place" as well as to "Denis Diderot and the 18th century encyclopedists." But the more immediate origins are in a project called *Nupedia* launched in March 2000 by Jimmy Wales and Larry Sanger. They were not the first to think of a free Web-based encyclopedia; in the earliest days of the Web, some had talked about creating a free "Interpedia"; in 1999 Richard Stallman, a key figure in the emergence of free and open-source software, proposed GNUpedia as a "Free Universal Encyclopedia and Learning Resource." The thirty-three-year-old Wales (also known as Jimbo), who got rich as an options trader and then became an Internet entrepreneur, decided to create a free, online encyclopedia. He recruited Sanger, age thirty-one, who was finishing a Ph.D. in philosophy at the Ohio State University—whom Wales knew from their joint participation in online mailing lists and Usenet discussion groups devoted to Ayn Rand and objectivism—to become the paid editor in chief. Wales's company Bomis, an Internet search portal and a vendor of online "erotic images" (featuring the Bomis Babe Report), picked up the tab initially.[6]

Sanger designed *Nupedia* to ensure that experts wrote and carefully vetted content. In part because of that extensive review, it managed to publish only about twenty articles in its first eighteen months. In early January 2001, as Sanger was trying to think of ways to make it easier for people without formal credentials to contribute to *Nupedia*, a computer programmer friend told him about the WikiWikiWeb software, developed by the programmer Ward Cunningham in the mid-1990s, that makes it easy to create or edit a Web page—no coding HTML (hypertext markup language) or uploading to a server needed. (Cunningham took the name from the Hawaiian word *wiki-wiki*, meaning "quick" or "informal.") Sanger thought that wiki users would quickly and informally create content for *Nupedia* that his experts would edit and approve. But the *Nupedia* editors viewed the experiment with suspicion; by mid-January Sanger and Wales had given it a separate name, *Wikipedia*, and its own domain.[7]

Very swiftly, *Wikipedia* became the tail that swallowed the dog (*Nupedia*). In less than a month, it had 1,000 articles; by the end of its first year, it had 20,000; by the end of its second year, it had 100,000 articles in just the English edition. (By then it had begun to spawn foreign-language editions, of which there are now 185, from Abkhazian to Klingon to Zulu, with the German edition the largest after English.) Sanger himself did not stay around to enjoy *Wikipedia*'s runaway growth. By late 2001 the tech boom was over, and Bomis, like most other dot-coms, was losing money and laying off employees. An effort to sell ads to pay Sanger's salary foundered as Internet advertising tanked, and Sanger lost his job in February 2002. He continued intermittently as a volunteer but finally broke with the project in January 2003 over the project's tolerance of problem participants and its hostility to experts.[8]

Since then, *Wikipedia*'s growth has accelerated. It had almost a half million articles by its third anniversary in January 2004; it broke the million mark just nine months later. More than fifty-five thousand people have made at least ten contributions to *Wikipedia*.[9] Over this short history, it has also evolved a style of operation and a set of operating principles that require explanation before any discussion of history on *Wikipedia*.

The *Wikipedia* Way: How It Works

The *Wikipedia* "Policies and Guidelines" page links to dozens of other pages, including six pages of "General Guidelines" (for example, "Contribute what you know or are willing to learn about"); twelve of "Behavior Guidelines" ("Don't bite the newcomers"); nineteen of "Content Guidelines" ("Check your facts"); nine of "Style Guidelines" ("Avoid one-sentence paragraphs"); and five of "Conventions" ("How to title articles"). But realizing that "they" (I employ the pronoun to refer to the collectivity of *Wikipedia* authors, editors, administrators, and programmers) would have no participants if authors were required to master this massive set of instructions before writing, they helpfully add, "You don't need to read every Wikipedia policy before you contribute!" and they offer a short primer of four "key policies."[10]

"*Wikipedia*," they declare first, "is an encyclopedia. Its goals go no further." Personal essays, dictionary entries, critical reviews, "propaganda or advocacy," and "original research" are excluded. Historians may find the last exclusion surprising since we value original research above everything else, but it makes sense for a collaboratively created encyclopedia. How can the collectivity assess the validity of statements if there is no verification beyond the claim "I discovered this in my research"?[11] As a result, *Wikipedia* (like encyclopedias in general) summarizes and reports the conventional and accepted wisdom on a topic but does not break new ground. And someone whose expertise rests on having done extensive original research on a topic gets no particular respect. That denigration of expertise contributed to Larry Sanger's split from the project.

The second key Wikipedian injunction is to "avoid bias." "Articles should be written from a neutral point of view [NPOV]," they insist, "representing differing views on a subject factually and objectively." Historians who learned (or teach) the mantra that "there is no objective history" in their undergraduate history methods class will regard that advice with suspicion. But Wikipedians quickly point out that the NPOV policy (as it is incessantly referred to in *Wikipedia* discussions) "doesn't assume that writing an article from a single, unbiased, objective point of view is possible." Instead, Wikipedians say they want to describe disputes rather than to take sides in them, to characterize differing positions fairly.[12]

Of course, writing "without bias"—even in the circumscribed way that *Wikipedia* defines it—is, as Wikipedians concede, "difficult" since "all articles are edited by people" and "people are inherently biased." But even if "neutrality" is a myth, it is a "founding myth" for *Wikipedia* much as "objectivity," according to Peter Novick, is a "founding myth" for the historical profession. *Wikipedia* articles rarely ascend to the desired level of neutrality, but the NPOV policy provides a shared basis of discourse among Wikipedians. On the "Discussion" pages that accompany every *Wikipedia* article, the number one topic of debate is whether the article adheres to the NPOV. Sometimes, those debates can go on at mind-numbing length, such as the literally hundreds of pages devoted to an entry on the Armenian genocide that still carries a warning that "the neutrality of this article is disputed."[13] *Wikipedia* entries on such controversial topics rarely succeed in meeting founder Jimmy Wales's goal of presenting "ideas and facts in such a fashion that both supporters and opponents can agree." But they surprisingly often achieve "a type of writing that is agreeable to essentially rational people who may differ on particular points." Unfortunately, that "type of writing" sometimes leads to mushy prose, exemplified by this description of the historian Daniel Pipes: "He is a controversial figure, both praised and condemned by other commentators."[14]

The third "key policy" is simpler: "don't infringe copyrights." Just as students can easily copy *Wikipedia* entries and submit them as term papers, *Wikipedia* authors can easily post prose copied from the vast plagiarism machine of the Web. But search engines make it relatively easy to catch both forms of plagiarism, and it does not seem to be much of a problem in *Wikipedia*. The more profound departure comes in the next sentence: *Wikipedia* "is a free encyclopedia licensed under the terms of the GNU Free Documentation License" (GFDL), a counterpart to the GNU General Public License (GPL) (used in free software projects such as Linux) designed for such open content as manuals and textbooks.[15]

The GFDL (and GPL) deviate most surprisingly from conventional intellectual property rules by giving you the freedom to use the text however you wish. As the license states: "You may copy and distribute the Document in any medium, either commercially or noncommercially, provided . . . you add no other conditions whatsoever to those of this License."[16] The "provided" clause means that any derivative document must inherit the same freedoms offered by the original—what GNUniks call "copyleft." You can publish a compilation of presidential biographies based on the profiles in *Wikipedia*; you can even rewrite half of them. But your new version must give credit to *Wikipedia* and allow others to reuse and refashion your revised version. In fact, multiple versions of *Wikipedia* content have sprouted all over the Web.

One further implication of *Wikipedia*'s implementation of free and open-source software principles is that its content is available to be downloaded, manipulated, and "data mined"—something not possible even with many resources (newspapers, for example) that can be read free online. *Wikipedia* can therefore be used for other purposes, including such questions-answering services as the Center for History and New Media's automated historical fact finder, H-Bot. Or it might provide the basis for tools that would enable you to search intelligently through quantities of undifferentiated digital text and distinguish, say, between references to John D. Rockefeller and those to his son John D. Rockefeller Jr. As Daniel J. Cohen has argued, resources such as *Wikipedia* "that are free to use in any way, even if they are imperfect, are more valuable than those that are gated or use-restricted, even if those resources are qualitatively better." Your freedom both to rewrite *Wikipedia* entries and to manipulate them for other purposes is thus arguably more profound than your ability to read them "for free." It is why free-software advocates say that to understand the concept of free software, you should think of "free speech" more than "free beer."[17]

The fourth pillar of *Wikipedia* wisdom is "respect other contributors."[18] Like writing without bias, it is easier said than done. What kind of respect, for example, do you owe a contributor who defaces other contributions or attacks other contributors? How do you ensure that entries are not continually filled with slurs and vandalism when the wiki allows any person anyplace to write whatever he or she pleases in any *Wikipedia* entry?

Wikipedia got by initially with a minimum of rules, in part to encourage participation.

We began [recalled Sanger] with no (or few) policies in particular and said that the community would determine—through a sort of vague consensus, based on its experience working together—what the policies would be. The very first entry on a "rules to consider" page was the "Ignore All Rules" rule (to wit: "If rules make you nervous and depressed, and not desirous of participating in the wiki, then ignore them entirely and go about your business").

Over time, however, rules proliferated. But *Wikipedia* acquired laws before it had police or courts. Sanger and Wales "agreed early on that, at least in the beginning, [they] should not eject anyone from the project except perhaps in the most extreme cases . . . despite the presence of difficult characters from nearly the beginning of the project." Sanger himself became increasingly distressed by the tolerance of "difficult people," or "trolls," on *Wikipedia*, believing they drove away "many better, more valuable contributors." Ultimately, the trolls wore Sanger down and pushed him out of the project.[19]

Although Sanger lost this battle, he may have won the war. *Wikipedia* gradually developed elaborate mechanisms for dealing with difficult people. It evolved intricate rules by which participants could be temporarily or even permanently banned from *Wikipedia* for inappropriate behavior. It also set up an elaborate structure of "administrators," "bureaucrats," "stewards," "developers," and elected trustees to oversee the project.[20] But the ideal remained to reach consensus—somewhat in the style of 1960s participatory democracy—rather than to impose formal discipline.

Standing over this noisy democratic polis, however, is the founder, Jimmy Wales—the "God-King," as some call him. The "banning policy" explains how users can be banned from *Wikipedia* by the "arbitration committee" or by Wikipedians acting "according to appropriate community-designed policies with consensus support." But it also adds tersely: "Jimbo Wales retains the power to ban users, and has used it." Wales's power rests not just on his prestige as founder but also on his place in the encyclopedia's legal structure. The Wikimedia Foundation, which controls *Wikipedia*, has a five-member board: two elected members plus Wales and two of his business partners.[21]

All of this works surprisingly well. To be sure, *Wikipedia* can be a bewildering and annoying place for newcomers. One familiar complaint is that " 'fanatic,' even 'kooky' contributors with idiosyncratic, out-of-mainstream, non-scientific belief systems can easily push their point of view, because nobody has the time and energy to fight them, and because they may be highly-placed in the Wikipedian bureaucracy." Yet somehow thousands of dispersed volunteers who do not know each other have organized a massive enterprise. Consensus and democracy fail at times. The Wikipedian collectivity must temporarily "lock" controversial entries because of vandalism and "edit wars" in which articles are changed and immediately changed back, such as an effort by NYCExpat to remove any references to Father Charles Coughlin's anti-Semitism. But other entries—even ones in which dedicated partisans such as the followers of Lyndon LaRouche battle for their point of view—remain open for anyone to edit and still present a reasonably accurate account.[22]

Wikipedia as History

Wikipedia has created a working community, but has it created a good historical resource? Are Wikipedians good historians? As in the old tale of the blind men and the elephant, your assessment of *Wikipedia* as history depends a great deal on what part you touch. It also depends, as we shall see, on how you define "history."

American historians might look first at the *Wikipedia* page headed "List of United States History Articles," which includes twelve articles surveying American history in conventional time periods and another thirty or so articles on such key topics as immigration, diplomatic history, and women's history. Unfortunately, the blind man reporting from those nether regions would return shaking his head in annoyance. He might start by complaining that the essay on the United States from 1918 to 1945 inaccurately describes the National Industrial Recovery Act of 1933 as in part a response to the "dissident challenges" of Huey Long and Father Charles Coughlin—a curious characterization of a law enacted when Coughlin was still an enthusiastic backer of Roosevelt and Long was an official (if increasingly critical) ally. But he would be much more distressed by the essay's incomplete, almost capricious, coverage than by the minor errors. Dozens of standard topics—the Red Scare, the Ku Klux Klan, the Harlem Renaissance, woman suffrage, the rise of radio, the emergence of industrial unionism—go unmentioned. And he would grind his teeth over the awkward prose and slack analysis ("the mood of the nation rejected Wilson's brand of internationalism") and the sometimes confusing structure (the paragraph on legislation passed in 1935 appears in the section on Roosevelt's second term).[23]

Other entries in the United States history series are worse. The entry on women leaves out the Nineteenth Amendment but devotes a paragraph to splits in the National Organization for Women (NOW) over the defense of

Valerie Solanas (who shot Andy Warhol). The 1865 to 1918 entry only briefly alludes to the Spanish-American War but devotes five paragraphs to the Philippine War, an odd reversal of the general bias in history books, which tend to ignore the latter and lavish attention on the former. The essay also plagiarizes one sentence from another online source. The 4,000-word essay on the history of U.S. immigration verges on incoherence and mentions famine-era Irish immigration only in a one-line picture caption.[24]

Part of the problem is that such broad synthetic writing is not easily done collaboratively. Equally important, some articles do not seem to have attracted much interest from Wikipedians. The essay on the interwar years has had only 137 edits, about one-seventh the number of interventions in the article on FDR. Participation in *Wikipedia* entries generally maps popular, rather than academic, interests in history. U.S. cultural history, recently one of the liveliest areas of professional history writing, is what *Wikipedia* calls a "stub" consisting of one banal sentence ("The cultural history of the United States is a broad topic, covering or having influence in many of the world's cultural aspects."). By contrast, *Wikipedia* offers a detailed 3,100-word article titled "Postage Stamps and Postal History of the United States," a topic with a devoted popular following that attracts little scholarly interest.[25]

Biographies of historical figures offer a more favorable terrain for *Wikipedia* since biography is always an area of popular historical interest. Moreover, biographies offer the opportunity for more systematic comparison because the unit of analysis is clear-cut, whereas other topics can be sliced and diced in multiple ways. But even to assess the quality of biographical writing in *Wikipedia* requires some context. You cannot compare, for example, *Wikipedia*'s 5,000 words on Martin Luther King Jr. with Taylor Branch's three-volume (2,900-page) prizewinning biography.[26] But how does it stack up against other reference works?

I judged 25 *Wikipedia* biographies against comparable entries in *Encarta*, Microsoft's well-regarded online encyclopedia (one of the few commercial encyclopedias that survive from a once-crowded marketplace), and in *American National Biography Online*, a high-quality specialized reference work published by Oxford University Press for the American Council of Learned Societies, written largely by professional historians, and supported by major grants. The comparison is unfair—both publications have had multimillion-dollar budgets—but it is still illuminating, and it sheds some favorable light on *Wikipedia*.[27]

In coverage *Wikipedia* currently lags behind the comprehensive *American National Biography Online*, which has 18,000 entries, but exceeds the general-interest *Encarta*. Of a sample of 52 people listed in *American National Biography Online*, *Wikipedia* included one-half, but *Encarta* only about one-fifth. The *American National Biography Online* profiles were also more detailed, averaging about four times as many words as those in *Wikipedia*. *Encarta* was the least detailed, with its entries for the sample only about one-quarter the length of *Wikipedia*'s.[28] Yet what is most impressive is that *Wikipedia* has found unpaid volunteers to write surprisingly detailed and reliable portraits of relatively obscure historical figures—for example, 900 words on the Union general Romeyn B. Ayres.

Relying on volunteers and eschewing strong editorial control leads to widely varying article lengths in *Wikipedia*. It devotes 3,500 words to the science fiction writer Isaac Asimov, more than it gives to President Woodrow Wilson (3,200) but fewer than it devotes to the conspiracy theorist and perennial presidential candidate Lyndon LaRouche (5,400); *American National Biography Online* provides a more proportionate (from a conventional historical perspective) coverage of 1,900 words for Asimov and 7,800 for Wilson. (It ignores the still-living LaRouche.) Of course, *American National Biography Online* also betrays the biases of its editors in its word allocations: Would nonhistorians agree that Charles Beard deserves twice as many words as the reformer and New Deal administrator Harold Ickes?

As the attention devoted to Asimov hints, *Wikipedia*'s authors do not come from a cross-section of the world's population. They are more likely to be English-speaking, males, and denizens of the Internet. Such bias has occasioned much discussion, including among Wikipedians. A page of candid self-criticism titled "Why Wikipedia Is Not So Great," acknowledges that "geek priorities" have shaped the encyclopedia: "There are many long and well-written articles on obscure characters in science fiction/fantasy and very specialised issues in computer science, physics and math; there are stubs, or bot [machine=generated] articles, or nothing, for vast areas of art, history, literature, film, geography." One regular contributor to *Wikipedia*'s history articles observed (somewhat tongue in cheek): "Wikipedia kicks Britannica's ass when it comes to online MMP [massively multiplayer] games, trading card games, Tolkieana and Star Wars factoids!" "This is the encyclopedia that Slashdot built" goes a familiar complaint that alludes to the early promotion of *Wikipedia* by the Web site that bills itself as the home of "news for nerds."

The "Google effect" further encouraged participation by Web surfers. As Sanger later explained, "each time Google spidered [crawled] the website, more pages would be indexed; the greater the number of pages indexed, the more people arrived at the project; the more people involved in the project, the more pages there were to index."[29]

Encyclopedia Britannica editor in chief Dale Hoiberg defensively pointed out to the *Guardian* that "Wikipedia authors write of things they're interested in, and so many subjects don't get covered; and news events get covered in great detail. The entry on Hurricane Frances is five times the length of that on Chinese art, and the entry on the British television show *Coronation Street* is twice as long as the article on Tony Blair." (Wikipedians responded to this criticism defensively, making the Blair entry 50 percent longer than the one on the television show.) But the largest bias—at least in the English-language version—favors Western culture (and English-speaking nations), rather than geek or popular culture.[30]

Perhaps as a result, *Wikipedia* is surprisingly accurate in reporting names, dates, and events in U.S. history. In the 25 biographies I read closely, I found clear-cut factual errors in only 4. Most were small and inconsequential. Frederick Law Olmsted is said to have managed the Mariposa mining estate after the Civil War, rather than in 1863. And some errors simply repeat widely held but inaccurate beliefs, such as that Haym Salomon personally loaned hundreds of thousands of dollars to the American government during the Revolution and was never repaid. (In fact, the money merely passed through his bank accounts.) Both *Encarta* and the *Encyclopedia Britannica* offer up the same myth.[31] The 10,000-word essay on Franklin Roosevelt was the only one with multiple errors. Again, some are small or widely accepted, such as the false claim (made by Roosevelt supporters during the 1932 election) that FDR wrote the Haitian constitution or that Roosevelt money was crucial to his first election to public office in 1910. But two are more significant—the suggestion that a switch by Al Smith's (rather than John Nance Garner's) delegates gave Roosevelt the 1932 nomination and the statement that the Supreme Court overruled the National Industrial Recovery Act (NIRA) in 1937, rather than 1935.

The lack of a single author or an overall editor means that *Wikipedia* sometimes gets things wrong in one place and right in another. The Olmsted entry has him (correctly) forming Olmsted, Vaux and Company in 1865 at the same time that he is (incorrectly) in California running Mariposa. The entry on

Andrew Jackson Downing says that Olmsted and Calvert Vaux designed Central Park in 1853 even though the cross-referenced article on Vaux has them (accurately) winning the design competition in 1858.[32]

To find four entries with errors in twenty-five biographies may seem a source for concern, but in fact it is exceptionally difficult to get every fact correct in reference works. "People don't realize how hard it is to nail the simplest things," noted Lars Mahinske, a senior researcher for *Britannica*. I checked ten *Encarta* biographies for figures that also appear in *Wikipedia*, and in the commercial product I found at least three biographies with factual mistakes. Even the carefully edited *American National Biography Online*, whose biographies are written by experts, contains at least one factual error in the twenty-five entries I examined closely, the date of Nobel Prize winner I. I. Rabi's doctoral degree—a date that *Wikipedia* gets right. Indeed, Wikipedians, who are fond of pointing out that respected reference sources have mistakes, gleefully publish a page devoted to "Errors in the *Encyclopedia Britannica* That Have Been Corrected in *Wikipedia*."[33]

Wikipedia, then, beats *Encarta* but not *American National Biography Online* in coverage and roughly matches *Encarta* in accuracy. This general conclusion is supported by studies comparing *Wikipedia* to other major encyclopedias. In 2004 a German computing magazine had experts compare articles in 22 different fields in the three leading German-language digital encyclopedias. It rated *Wikipedia* first with a 3.6 on a 5-point scale, placing it above *Brockhaus Premium* (3.3) and *Encarta* (3.1). The following year the British scientific magazine *Nature* asked experts to assess 42 science entries in *Wikipedia* and *Encyclopedia Britannica*, without telling them which articles came from which publication. The reviewers found only 8 serious errors, such as misinterpretations of major concepts—an equal number in each encyclopedia. But they also noted that *Wikipedia* had a slightly larger number (162 versus 123) of smaller mistakes, including "factual errors, omissions or misleading statements." *Nature* concluded that "*Britannica*'s advantage may not be great, at least when it comes to science articles," and that "considering how *Wikipedia* articles are written, that result might seem surprising."[34]

Thus, the free and open-source encyclopedia *Wikipedia* offers a formidable challenge to the well-established and seemingly authoritative *Encyclopedia Britannica* as well as to Microsoft's newer and well-regarded *Encarta* just as the free and open-source Linux operating system now seriously challenges Microsoft's Windows in the server market. Not surprisingly, *Encarta* has been scrambling to compete—both by making its content more generally available (you can get free access by using the MSN search engine) and by inviting readers to propose edits to the content.

If the unpaid amateurs at *Wikipedia* have managed to outstrip an expensively produced reference work such as *Encarta* and provide a surprisingly comprehensive and largely accurate portrait of major and minor figures in U.S. history, professional historians need not fear that Wikipedians will quickly put them out of business. Good historical writing requires not just factual accuracy but also a command of the scholarly literature, persuasive analysis and interpretations, and clear and engaging prose. By those measures, *American National Biography Online* easily outdistances *Wikipedia*.

Compare, for example, *Wikipedia*'s 7,650-word portrait of Abraham Lincoln with the 11,000-word article in *American National Biography Online*. Both avoid factual errors and cover almost every important episode in Lincoln's life. But surely any academic would prefer the *American National Biography Online* sketch by the prominent Civil War historian James McPherson. Part of the difference lies in McPherson's richer contextualization (such as the concise explanation of the rise of the Whig party) and his linking of Lincoln's life to dominant themes in the historiography (such as free-labor ideology). But McPherson's profile is distinguished even more by his artful use of quotations to capture Lincoln's voice, by his evocative word portraits (the young Lincoln was "six feet four inches tall with a lanky, rawboned look, unruly coarse black hair, a gregarious personality, and a penchant for telling humorous stories"), and by his ability to convey a profound message in a handful of words ("The republic endured and slavery perished. That is Lincoln's legacy"). By contrast, *Wikipedia*'s assessment is both verbose and dull: "Lincoln's death made the President a martyr to many. Today he is perhaps America's second most famous and beloved President after George Washington. Repeated polls of historians have ranked Lincoln as among the greatest presidents in U.S. history."[35]

In addition to McPherson's elegant prose, his profile embodies the skill and confident judgment of a seasoned historian. The same is true of many other *American National Biography Online* sketches—Alan Brinkley on Franklin Roosevelt or T. H. Watkins on Harold Ickes, for example. Those gems of short biographical writing combine crisp prose with concise judgments about the significance of their subjects. Even less masterly entries in *American National Biography Online* generally sport smoother prose than *Wikipedia*. And they also offer reliable bibliographic essays with the latest scholarly works. *Wikipedia* entries generally include references, but not always the best ones. The bibliography for Haym Salomon contains only two works, both published more than fifty years ago. Of one of those books, *American National Biography Online* warns that it "repeats all the myths and fabrications found in earlier accounts."[36]

Of course, not all historians write as well as McPherson and Brinkley, and some of the better-written *Wikipedia* entries provide more engaging portraits than some sterile and routine entries in *American National Biography Online*. For example, the *American National Biography Online* sketch of the Hall of Fame pitcher Red Faber provides a plodding, almost year-by-year account, whereas *Wikipedia* gives a more concise overview of his career and significance. *Wikipedia*'s profile of the Confederate guerrilla fighter William Clarke Quantrill arguably does a better job of detailing the controversies about his actions than *American National Biography Online*. Even so, it provides a typical waffling conclusion that contrasts sharply with the firm judgments in the best of the *American National Biography Online* essays: "Some historians," they write, "remember him as an opportunistic, bloodthirsty outlaw, while others continue to view him as a daring soldier and local folk hero."[37]

This waffling—encouraged by the NPOV policy—means that it is hard to discern any overall interpretive stance in *Wikipedia* history. One might expect—given the Randian politics of the founders and the strength of libertarian sentiments in cyberspace—a libertarian or conservative slant. But I did not find it. One can see occasional glimmers, as in the biography of Calvin Coolidge that says with apparent approval, "Coolidge was the last President of the United States who did not attempt to intervene in free markets, letting business cycles run their course." This sentence was inserted early on by an avowed libertarian and it has survived dozens of subsequent edits. But *Wikipedia* also presents the socialist Eugene V. Debs in flattering terms; the only criticism is that he "underestimated the lasting power of racism." At least one conservative blogger charges that *Wikipedia* is "more liberal than the liberal media."[38]

If anything, the bias in *Wikipedia* articles favors the subject at hand. "Articles tend to be whatever-centric," they acknowledge in one of their many self-critical commentaries. "People point out whatever is exceptional about their home province, tiny town or bizarre hobby, without noting frankly that their home province is completely unremarkable, their tiny town is not really all that special or that their bizarre hobby is, in fact, bizarre." That localism can sometimes cause conflicts on nonlocal entries, as in the Olmsted profile, where a Wikipedian from Louisville complains on the "Discussion" page that the biography overestimates Olmsted's work in Buffalo and ignores his work in—surprise!—Louisville.[39]

Moreover, the collective mode of composition in *Wikipedia* and the repeated invocation of the NPOV policy mean that it tends to avoid controversial stands of all kinds. Whereas there is much popular interest in lurid aspects of history, *Wikipedia* editors shy away from sensationalist interpretations (although not from discussion of controversies about such interpretations). The biography of Warren G. Harding cautiously warns of "innuendo" and "speculation" surrounding his extramarital affairs, expresses doubt about his alleged affair with Nan Britton, and insists that there is "no scientific or legal basis" for the rumors of Harding's mixed "blood." And while popular history leans toward conspiracy theories, *Wikipedia* seems more likely to debunk them. It judiciously concludes that there is "no evidence" that Roosevelt "knew all about the planned attack on Pearl Harbor but did nothing to prevent it."[40]

Overall, writing is the Achilles' heel of *Wikipedia*. Committees rarely write well, and *Wikipedia* entries often have a choppy quality that results from the stringing together of sentences or paragraphs written by different people. Some Wikipedians contribute their services as editors and polish the prose of different articles. But they seem less numerous than other types of volunteers. Few truly gifted writers volunteer for *Wikipedia*. *Encarta*, while less comprehensive than *Wikipedia*, generally offers better—especially, more concise—writing.

Even so, few would turn to *Encarta* or the *Encyclopedia Britannica* for good writing. Like other such works, *Wikipedia* employs the "encyclopedia voice," a product, the former *Encyclopedia Britannica* editor Robert McHenry argued, of "a standardized process and standardized forms, and . . . a permanent editorial staff, whose members train their successors in what amounts to an apprenticeship." It also reflects reference works' general allergy to strongly stated opinions. More than forty years ago, Charles Van Doren, who became a senior editor at *Encyclopedia Britannica* after his quiz show debacle, complained that "the tone of American encyclopedias is often fiercely inhuman. It appears to be the wish of some contributors to write about living institutions as if they were pickled frogs, outstretched upon a dissecting board." Contrast any modern encyclopedia entry with this one on John Keats by Algernon Charles Swinburne, in the (late nineteenth-century) ninth edition of the *Encyclopedia Britannica*: "The Ode to a Nightingale, one of the final masterpieces of human work in all time and for all ages, is immediately preceded in all editions now current by some of the most vulgar and fulsome doggerel ever whimpered by a vapid and effeminate rhymester in the sickly stage of whelphood."[41]

Swinburne's "bias" would have transgressed not only *Wikipedia*'s NPOV but also the preference of conventional, modern encyclopedias for what McHenry calls "the blandness of mere information." Indeed, the NPOV mimics conventional "encyclopedia style." "Wikipedia users," two social scientists conclude, "appropriate norms and expectations about what an 'encyclopedia' should be, including norms of formality, neutrality, and consistency, from the larger culture." As a result, they find, over time *Wikipedia* entries become "largely indistinguishable stylistically from [those in] the expert-created *Columbia Encyclopedia*."[42]

Conversely, the worst-written entries are the newest and least edited. As the "Replies to Common Objections" page explains: "Wikipedia has a fair bit of well-meaning, but ill-informed and amateurish work. In fact, we welcome it—an amateurish article to be improved later is better than nothing."[43] That means you can encounter both the polished entry on Red Faber and the half-written article on women's history. Less sophisticated readers may not know the difference.

They also may not realize when an article has been vandalized. But vandalism turns out to be less common than one would expect in a totally open system. Over a two-year period, vandals defaced the Calvin Coolidge entry only ten times—almost all with obscenities or juvenile jottings that would have not misled any visitor to the site. (The one exception changed his birth date to 1722, which was also unlikely to confuse anyone.) The median time for repairing the damage was three minutes.[44] More systematic tests have found that vandalism generally has a short life on *Wikipedia*. The blogger Alex Halavais, graduate director for the informatics school at the University at Buffalo, inserted thirteen small errors into *Wikipedia* entries—including, for example, the claim that the "well-known abolitionist Frederick Douglass made Syracuse his home for four years." To his surprise, vigilant Wikipedians removed all the mistakes within two and a half hours. Others have been more successful in slipping errors into the encyclopedia, including an invented history of Chesapeake, Virginia, describing it as a major importer of cow dung until "it collapsed in one tremendous heap," which lasted on *Wikipedia* for a month.[45] But vandals face formidable countermeasures that *Wikipedia* has evolved over time, including a "recent changes patrol" that constantly monitors changes reported on a "Recent Changes" page as well as "personal watchlists" that tell contributors whether an article of interest to them has been changed. On average, every article is on the watchlist of two accounts, and the keepers of those lists often obsessively check them several times a day. More generally, the sheer volume of edits—almost 100,000 per day—means that entries, at least popular entries, come under almost constant scrutiny.[46]

But, as a fall 2005 controversy involving an entry on the journalist John Seigenthaler makes clear, *Wikipedia*'s controls and countermeasures are a work in progress, and vandalism in infrequently read entries can slip under the radar. In May 2005 Brian Chase altered the article on Seigenthaler to play a "joke" on a co-worker at Rush Delivery in Nashville, Tennessee, where Seigenthaler's late brother had been a client. The not very humorous change suggested that Seigenthaler, who once worked for Robert Kennedy, was thought "to have been directly involved in the Kennedy assassinations of both John, and his brother, Bobby." In September, Seigenthaler learned about the scurrilous charges and complained to Jimmy Wales, who removed them from both the active page and the page history. But, as Seigenthaler wrote in *USA Today* in late November, "the false, malicious 'biography'" had "appeared under [his] name for 132 days." Moreover, sites that mirror *Wikipedia*'s content such as Answers.com and Reference.com retained the falsehoods for another three weeks. The episode received wide notice, with many *Wikipedia* critics echoing Seigenthaler's charge that the online encyclopedia "is a flawed and irresponsible research tool" where "volunteer vandals with poison-pen intellects" abound.[47]

Wikipedia's defenders complained, in the words of Paul Saffo, director of the Institute for the Future, that Seigenthaler "clearly doesn't understand the culture of Wikipedia." Saffo and others argued that Seigenthaler "should have just changed" the false statements. But Seigenthaler pointed out that the lies were online for several months before he even knew about them and that he did not want to have anything to do with the flawed enterprise. A more persuasive defense, offered by others, acknowledged the flaws but pointed out the relative ease of correcting them. After all, malicious gossip has long surrounded public figures, but it is very hard to track down and stop. Even when it appears in print publications, which are subject (as *Wikipedia* is not) to libel laws, the only remedy is going to court. In the case of *Wikipedia*, the defamatory statements about Seigenthaler were entirely expunged. Professor Lawrence Lessig of the Stanford Law School argued that defamation is a by-product of free speech and that while "Wikipedia is not immune from that kind of maliciousness . . . it is, relative to other features of life, more easily corrected." As Wade Roush, an editor at *TechnologyReview. com*, wrote in his blog, "the community-editing model gives us a newfound power to create wrongs—but also to reverse wrongs."[48]

Still, the episode eroded *Wikipedia*'s credibility and led to efforts at damage control. Jimmy Wales announced that *Wikipedia* would now require users to register before creating new articles. Of course, that rule would not have stopped Brian Chase because registration will not be required simply to edit an existing entry. Moreover, registration may actually provide less accountability; you need not report even an e-mail address to register, whereas unregistered users have their Internet Protocol (IP) addresses recorded, and it was such an address that made it possible to track down Chase. And *Wikipedia* still lacks any mechanism for guaranteeing an entry's accuracy at the moment when you land on the site; a vandal or even a scholar trying to test the system might have just changed the "fact" that you are seeking. Wikipedians have discussed possible solutions to this problem. For example, visitors could have the option of viewing only a version of an article that had been "patrolled," that is, checked for random vandalism, or users could have the choice of seeing an "approved" page or one "pending" approval from a certain number of editors.[49]

Wikipedia already offers a limited version of that choice by allowing you to check the page's "history." The wiki software allows you to compare every single version of an article going back to its creation. In a widely circulated critique of *Wikipedia*, the former *Encyclopedia Britannica* editor McHenry observed that "the user who visits Wikipedia . . . is rather in the position of a visitor to a public restroom. It may be obviously dirty, so that he knows to exercise great care, or it may seem fairly clean, so that he may be lulled into a false sense of security. What he certainly does not know is who has used the facilities before him." McHenry is right about the "publicness" of *Wikipedia*, but why not choose a more uplifting analogy, like the public school or the public park? Moreover, he is wrong about not knowing what came before you. The "History" page tells you not only who used the facilities (at least their usernames or IP addresses) but also precisely what they did there. Indeed, simply taking information buried on the "History" page and making it more public would enhance *Wikipedia*—for example, the "Article" page might say, "This article has been edited 350 times since it was created on May 5, 2002, including 30 times in the past week." It could even add that "very active Wikipedians" (those with more than 100 edits this month) contributed 52 percent of those edits. Such information could be automatically generated, and it would give the reader additional clues to the quality of the entry. Another possible improvement would have readers rate the quality of individual *Wikipedia* entries, an approach used by a number of popular Internet sites, including Amazon.com (which enjoins visitors not just to review and rate books but also to answer the question "Was this review helpful to you?") and *Slashdot* (which has a complex system of "moderation" that rates the quality of posted comments). During the Seigenthaler controversy, Wales announced that *Wikipedia* would be adding this feature soon.[50]

As Roush, Lessig, and others argued amid the Seigenthaler uproar, *Wikipedia*'s lack of fixity also has a more positive face—it can be updated instantly. Wikipedians like to point out that after the Indian Ocean tsunami of 2004 they added relevant entries within hours, including animations, geological information, reports on the international relief effort, and comprehensive links. Of course, the ability to capture the news of the day is of less interest to historians, but *Wikipedia* has also quickly captured the latest historical "news." You had to wait until the morning of June 1, 2005, to learn from your local newspaper that W. Mark Felt had been unmasked as "Deep Throat," but even before the evening news on May 31 you could have read about it in *Wikipedia*'s article on the "Watergate scandal." Like journalism, *Wikipedia* offers a first draft of history, but unlike journalism's draft, that history is subject to continuous revision. *Wikipedia*'s ease of revision not only makes it more up-to-date than a traditional encyclopedia, it also gives it (like the Web itself) a self-healing quality since defects that are criticized can be quickly remedied and alternative perspectives can be instantly added. McHenry's critique, for example, focused on problems in the entry on Alexander Hamilton. Two days later, they were fixed.[51]

Why Should We Care? Implications for Historians

One reason professional historians need to pay attention to *Wikipedia* is because our students do. A student contributor to an online discussion about *Wikipedia* noted that he used the online encyclopedia to study the historical terms for a test on early romanticism in Britain. Other students routinely list it in term paper bibliographies. We should not view this prospect with undue alarm. *Wikipedia* for the most part gets its facts right. (The student of British culture reported that *Wikipedia* proved as accurate as the *Encyclopedia Britannica* and easier to use.) And the general panic about students' use of Internet sources is overblown. You can find bad history in the library, and while much misinformation circulates on the Internet, it also helps to debunk myths and to correct misinformation.[52]

Yet the ubiquity and ease of use of *Wikipedia* still pose important challenges for history teachers. *Wikipedia* can act as a megaphone, amplifying the (sometimes incorrect) conventional wisdom. As Wikinfo (a fork, or spin-off, from *Wikipedia*) explains: "A wiki with so many hundreds of thousands of pages is bound to get some things wrong. The problem is, that because Wikipedia has become the 'AOL' [America Online] of the library and reference world, such false information and incorrect definitions of terms become multiple incompetences, propagated to millions of potential readers worldwide." Not only does *Wikipedia* propagate misinformation but so do those who appropriate its content, as they are entitled to do under the GFDL. As a result, as the blogger John Morse observed, "when you search Google for some obscure term that Wikipedia knows about, you might get two dozen results that all say the same thing—seemingly authoritative until you realize they all spread from a snapshot of Wiki—one that is now severed from the context of editability and might seem more creditable than it really is." The Web site Answers.com, which promises to provide "quick, integrated reference answers," relies heavily on *Wikipedia* for those answers. And Google, which already puts *Wikipedia* results high in its rankings, now sends people looking for "definitions" to Answers.com. Can you hear the sound of one hand clapping?[53]

Wikipedia's ease of use and its tendency to show up at the top of Google rankings in turn reinforce students' propensity to latch on to the first source they encounter rather than to weigh multiple sources of information. Teachers have little more to fear from students' *starting* with *Wikipedia* than from their starting with most other basic reference sources. They have a lot to fear if students *stop* there. To state the obvious: *Wikipedia* is an encyclopedia, and encyclopedias have intrinsic limits. Most academics have not relied heavily on encyclopedias since junior high school days. And most do not want their students to rely heavily on encyclopedias—digital or print, free or subscription, professionally written or amateur and collaborative—for research papers. One *Wikipedia* contributor noted that despite her "deep appreciation for it," she still "roll[s her] eyes whenever students submit papers with *Wikipedia* as a citation." "Any encyclopedia, of any kind," wrote another observer, "is a horrible place to get the whole story on any subject." Encyclopedias "give you the topline"; they are "the *Reader's Digest* of deep knowledge." Fifty years ago, the family encyclopedia provided this "rough and ready primer on some name or idea"; now that role is being played by the Internet and increasingly by *Wikipedia*.[54]

But should we blame *Wikipedia* for the appetite for predigested and prepared information or the tendency to believe that anything you read is true? That problem existed back in the days of the family encyclopedia. And one key solution remains the same: spend more time teaching about the limitations of all information sources, including *Wikipedia*, and emphasizing the skills of critical analysis of primary and secondary sources.

Another solution is to emulate the great democratic triumph of *Wikipedia*—its demonstration that people are eager for free and accessible information resources. If historians believe that what is available free on the Web is low quality, then we have a responsibility to make better information sources available online. Why are so many of our scholarly journals locked away behind subscription gates? What about *American National Biography Online*—written by professional historians, sponsored by our scholarly societies, and supported by millions of dollars in foundation and government grants? Why is it available only to libraries that often pay thousands of dollars per year rather than to everyone on the Web as *Wikipedia* is? Shouldn't professional historians join in the massive democratization of access to knowledge reflected by *Wikipedia* and the Web in general?[55] *American National Biography Online* may be a significantly better historical resource than *Wikipedia*, but its impact is much smaller because it is available to so few people.

The limited audience for subscription-based historical resources such as *American National Biography Online* becomes an even larger issue when we move outside the borders of the United States and especially into poorer parts of the world, where such subscription fees pose major problems even for libraries. Moreover, in some of those places, where censorship of textbooks and other historical resources is common, the fact that *Wikipedia's* freedom means both "free beer" and "free speech" has profound implications, because it allows the circulation of alternative historical voices and narratives. Some repressive governments have responded by restricting access to *Wikipedia*. China, for example, currently prevents its citizens from reading the English- or Chinese-language versions of *Wikipedia*. And it is probably not a coincidence that the first blocking of *Wikipedia* in China began on the fifteenth anniversary of the Tiananmen Square protests.[56]

Professional historians have things to learn not only from the open and democratic distribution model of *Wikipedia* but also from its open and democratic production model. Although *Wikipedia* as a product is problematic as a sole source of information, the process of creating *Wikipedia* fosters an appreciation of the very skills that historians try to teach. Despite *Wikipedia*'s unconventionality in the production and distribution of knowledge, its epistemological approach—exemplified by the NPOV policy—is highly conventional, even old-fashioned. The guidelines and advice documents that *Wikipedia* offers its editors sound very much like the standard manuals offered in undergraduate history methods classes. Editors are enjoined, for example, to "cite the source" and to check their facts and reminded that "verifiability" is an "official policy" of *Wikipedia*. An article directed at those writing articles about history for *Wikipedia* explains (in the manner of a History 101 instructor) the difference between primary and secondary sources and also suggests helpfully that "the correct standard of material to generate encyclopedic entries about historical subjects are: 1. Peer reviewed journal articles from a journal of history; 2. Monographs written by historians (BA Hons (Hist), MA, PhD); 3. Primary sources."[57]

Participants in the editing process also often learn a more complex lesson about history writing—namely that the "facts" of the past and the way those facts are arranged and reported are often highly contested. One *Wikipedia* guideline document reports with an air of discovery: "Although it doesn't seem to be logical to worry about a Wikipedia article, people do battle over history and the way it is written all the time." And such skirmishes break out all over *Wikipedia*. Each article contains a companion "Discussion" page, and on those pages, editors engage—often intensely—in what can only be called historiographic debate. Was Woodrow Wilson a racist? Did the New Deal resolve the problems of the Great Depression? Sometimes relatively narrow issues are debated (for example, William Jennings Bryan's role in the passage of the Butler Act, which prohibited the teaching of evolution in Tennessee) that open up much broader issues (for example, the sources of antievolution sentiment in the 1920s).[58]

Wikipedia has even developed its own form of peer review in its debates on whether articles deserve "featured article" status. Those aspiring to have their articles receive that status—given to the best .1 percent of articles as judged by such criteria as completeness, factual accuracy, and good writing—

are encouraged to request "peer review" in order to "expose articles to closer scrutiny than they might otherwise receive."[59] Then further public debate decides whether Wikipedians agree on awarding featured article status.

Thus, those who create *Wikipedia*'s articles and debate their contents are involved in an astonishingly intense and widespread process of democratic self-education. *Wikipedia*, observes one *Wikipedia* activist, "teaches both contributors and the readers. By empowering contributors to inform others, it gives them incentive to learn how to do so effectively, and how to write well and neutrally." The classicist James O'Donnell has argued that the benefit of *Wikipedia* may be greater for its active participants than for its readers: "A community that finds a way to talk in this way is creating education and online discourse at a higher level."[60]

My colleagues at the Center for History and New Media interviewed people who regularly contribute to history articles on *Wikipedia*, and a passion for self-education comes through in numerous interviews. A Canadian contributor, James Willys Rosenzweig (no relation), observed that his "involvement in *Wikipedia* [is] a natural fit" because "I am interested in a broad variety of subjects, and I read for pleasure in as many fields as I can." APWoolrich, a British contributor who left school at age sixteen and became an ardent self-taught industrial archeologist, answered the question "Why do I enjoy it?" with "It beats TV any day, in my view!"[61]

But APWoolrich is as enthusiastic about contributing to the education of others as to his own. *Wikipedia*, he told us, "accords with my personal philosophy of sharing knowledge, and it links me with the rest of humanity." He believes we have a "duty" to share knowledge "without thought of reward." "*Wikipedia* is the 'Invisible College' concept revived for the 21st century." A blind high school student had a different reference point. "It is almost like playing a computer game but it is actually useful because it helps someone anywhere in the world get information that is uncluttered by junk," he told us. "I think of myself as a teacher," said Einar Kvaran, an uncredentialed "art historian without portfolio," who spends about six hours a day writing articles about American art and sculpture. Like bloggers and amateur Web site developers, contributors to *Wikipedia* enjoy the opportunity to make their work public and to contribute to building the public space of the Web.[62]

Should those who write history for a living join such popular history makers in writing history in *Wikipedia*? My own tentative answer is yes.[63] If *Wikipedia* is becoming the family encyclopedia for the twenty-first century, historians probably have a professional obligation to make it as good as possible.

And if they devoted just one day to improving the entries in their areas of expertise, it would not only significantly raise the quality of *Wikipedia*, it would also enhance popular historical literacy. Historians could similarly play a role by participating in the populist peer review process that certifies contributions as featured articles.

Still, my view is tempered by the recognition that the encounter between professional historians and amateur Wikipedians is likely to be rocky at times. That seems to have been particularly true in the early days of *Wikipedia*. Larry Sanger reported that some of the earliest contributors were "academics and other highly-qualified people"—including two historians with Ph.D.s—who "were slowly worn down and driven away by having to deal with difficult people on the project." "I feel that my integrity has been questioned," the historian J. Hoffmann Kemp wrote in signing off in August 2002. "I'm too tired to play anymore."[64]

Even Jimmy Wales, who has been more tolerant of "difficult people" than Sanger, complained about "an unfortunate tendency of disrespect for history as a professional discipline." He saw the tendency reflected in historical entries that synthesize "work in a non-standard way" and "produce novel narratives and historical interpretations with citation to primary sources to back up their interpretation of events." He noted that "some who completely understand why *Wikipedia* ought not create novel theories of physics by citing the results of experiments and so on and synthesizing them into something new, may fail to see how the same thing applies to history."[65]

But the flip side of Wales's respect for the historical discipline, as expressed in the ban on original research (and original interpretations), is that it seemingly limits professional historians' role in *Wikipedia*. The "no original research policy" means that you cannot offer a startling new interpretation of Warren Harding based on newly uncovered sources. As a result, while *Wikipedia* officially "welcomes experts and academics," it also warns that "such experts do not occupy a privileged position within *Wikipedia*. They should refer to themselves and their publications in the third person and write from a neutral point of view (NPOV). They must also cite publications, and may not use their unpublished knowledge as a source of information (which would be impossible to verify)."[66]

Even a comparison that focuses on the ban on original research understates the differences between professionals and amateurs. For one thing, historical expertise does not reside primarily in the possession of some set of obscure facts. It relies more often on a deep acquaintance with a wide variety of already published narratives and an ability to synthesize those narratives (and facts) coherently. It is considerably easier to craft a policy about "verifiability" or even "neutrality" than about "historical significance." Professional historians might find an account accurate and fair but trivial; that is what some see as the difference between history and antiquarianism. Thus, the conflict between professionals and amateurs is not necessarily a simple one over whether people are doing good or bad history but a more complex (and more interesting) conflict about what kind of history is being done. Comparing the free *Wikipedia* and the costly and expensively produced *American National Biography Online* erects professional historical scholarship as a transhistorical and transcultural standard of history writing when we know that there are many ways of writing and talking about the past. What is particularly interesting and revealing about *Wikipedia* is its reflection of what we could call a "popular history poetics" that follows different rules from conventional professional scholarship.[67]

One noticeable difference is the affection for surprising, amusing, or curious details—something that *Wikipedia* shares with other forms of popular historical writing, such as articles in *American Heritage* magazine. Consider some details that Wikipedians include in their Lincoln biography that do not make their way into McPherson's profile: Lincoln's sharing a birthday with Charles Darwin; his nicknames (the Rail Splitter is mentioned twice); his edict making Thanksgiving a national holiday; and the end of his bloodline with the death of Robert Beckwith in 1985. Not surprisingly, *Wikipedia* devotes five times as much space to Lincoln's assassination as the longer *American National Biography Online* profile does.[68] The same predilection for colorful details marks other portraits. We learn from the Harding biography that the socialist Norman Thomas was a paper boy for the *Marion Daily Star* (which Harding owned), that Harding reached the sublime degree as a Master Mason, and that Al Jolson and Mary Pickford came to Marion, Ohio, during the 1920 campaign for photo ops. It devotes two paragraphs to speculation about whether Harding had "Negro blood" and five paragraphs to his extramarital affairs. Meanwhile, key topics—domestic and foreign policies, the Sheppard-Towner Maternity and Infancy Act of 1921, immigration restriction, and naval treaties—are ignored or hurried over. We similarly learn that Woodrow Wilson belonged to Phi Kappa Psi fraternity and wrote his initials on the underside of a table in the Johns Hopkins University history department, but not about his law practice or his intellectual development at Princeton University.[69]

Wikipedia's view of history is not only more anecdotal and colorful than professional history, it is also—again like much popular history—more factualist. That is reflected in the incessant arguing about NPOV, but it can also be seen in the obsession with list making. The profile of FDR leads you not just to a roll of all presidents but also to a list of every secretary of the interior, every chairman of the Democratic National Committee, every key event that happened on April 12 (when Roosevelt died), and every major birth in 1882 (when he was born). From the perspective of professional historians, the problem of Wikipedian history is not that it disregards the facts but that it elevates them above everything else and spends too much time and energy (in the manner of many collectors) on organizing those facts into categories and lists.

Finally, Wikipedian history is presentist in a slightly different way from that of professional history—where, for example, a conservative turn in the polity leads us to reevaluate conservatism in the past. Rather, *Wikipedia* entries often focus on topics that have ignited recent public, not just professional, controversy. The topic of Lincoln's sexuality—not mentioned by McPherson—occupied so much of the *Wikipedia* biography that in December 2004 a separate 1,160-word entry was created that focuses on C. A. Tripp's controversial, then-recent book *The Intimate World of Abraham Lincoln*. The entry on the Spanish-American War examines in considerable detail whether the *Maine* was sunk by a mine (a subject in the news as the result of a 1998 *National Geographic* study) but pays no attention to the important (to professional historians) arguments of Kristin L. Hoganson's book of the same year that "gender politics" provoked the war.[70]

That the latest article in *National Geographic* rather than the latest book from Yale University Press shapes *Wikipedia* entries reflects the fact that *Wikipedia* historians operate in a different world than historians employed in universities. Although *Wikipedia* enjoins its authors to "cite the source," that policy is honored mainly in the breach—unlike in academic historical journals, where authors and editors obsess over proper and full citation. Moreover, the bibliographies offered after *Wikipedia* entries are often incomplete or out-of-date—a cardinal sin in professional history. Yet Wikipedians are mindful of a wider community of "historians." It is just that for them the most important community is authors of other *Wikipedia* entries. And every article includes literally dozens of cross-references (links) to other *Wikipedia* articles.

An account of Lincoln's life that focuses on debates about his sexuality and dwells on his birth date, nicknames, and assassination is not "wrong," but it is not the kind of brief account that a professional historian such as McPherson would write. Professional historians who enter the terrain of *Wikipedia* will have an easy time correcting the year when the Supreme Court invalidated the NIRA but a much harder time eliminating Lincoln's nicknames. Wikipedians would agree with professional historians that the Supreme Court decision happened on a particular day, but they might not agree that Lincoln's nicknames are "unimportant" or "uninteresting." And such historians will have to decide how much of their disciplinary "authority" they are prepared to "share" in this new public space.[71]

Although making people we generally view as our audience into our collaborators may prove unsettling, it will also be instructive. One history doctoral student at an Ivy League institution who has contributed actively to *Wikipedia* explained that "I use it primarily to practice writing for a nonacademic audience, and as a way to solidify my understanding of topics (nothing helps one remember things like rewriting it)." He added, "I regard my Wikipedia contributions as informal and relatively anonymous, and use a much more casual demeanor than one would use in a professional setting (that is, I often tell people they don't know what they're talking about)."[72] If *Wikipedia* teaches us (and our students) to speak more clearly to the public and to say more clearly what is on our minds, it will have a positive impact on academic culture.

But a much broader question about academic culture is whether the methods and approaches that have proven so successful in *Wikipedia* can also affect how scholarly work is produced, shared, and debated. *Wikipedia* embodies an optimistic view of community and collaboration that already informs the best of the academic enterprise. The sociologist Robert K. Merton talked about "the communism of the scientific ethos," and communal sharing is an ideal that some historians hold and that many of our practices reflect, even while alternative, more individualistic and competitive, modes also thrive.[73]

Can the wiki way foster the collaborative creation of historical knowledge? One promising approach would leverage the volunteer labor of amateurs and enthusiasts to advance historical understanding. Historians have, of course, benefited from the labors of amateurs and volunteers. Think of the generations of local historians who have collected, preserved, and organized historical documents subsequently mined by professional historians. But the new technology of the Internet opens up the possibility of much more °massive efforts relying on what the legal scholar Yochai Benkler has called "commons-based peer production." The "central characteristic" of such production, wrote Benkler, "is that groups of individuals successfully collaborate on large-scale projects following a diverse cluster of motivational drives and social signals, rather than either market prices or managerial commands." "Ubiquitous computer communications networks," he argued, have brought about "a dramatic change in the scope, scale, and efficacy of peer production."[74] The most prominent recent example of such non-market based peer production is free and open-source software. The Internet would now grind to a halt without such free and open-source resources as the operating system Linux, the Web server software Apache, the database MySql, and the programming language PHP.

Yet, as Benkler showed, the peer production of information is much broader than free software, and he offers *Wikipedia* as one notable example. Another—and one perhaps more relevant to professional historians—is the National Aeronautics and Space Administration (NASA)'s Ames Clickworkers project, which encouraged volunteers to "mark craters on maps of Mars, classify craters that have already been marked, or search the landscape of Mars for 'honeycomb' terrain." In six months, more than 85,000 people visited the site and made almost 2 million entries. An analysis of the markings found that "the automatically-computed consensus of a large number of clickworkers is virtually indistinguishable from the inputs of a geologist with years of experience in identifying Mars craters."[75]

Probably the closest historical equivalent to the NASA clickworkers are the legions of volunteer genealogists who have been digitizing thousands of documents. For example, volunteers working for the Church of Jesus Christ of Latter-day Saints digitized the records of the 55 million people listed in the 1880 United States census and the 1881 Canadian census and made them available for free at the church's FamilySearch Internet Genealogy Service. Another volunteer effort, Project Gutenberg, has created an online repository of 15,000 e-texts of public domain books. Optical character recognition (OCR) software can relatively cheaply and automatically digitize print works, but it is generally only 95–99 percent accurate. To get a fully clean text is more expensive. Enter "distributed proofreaders"—a collaborative Web-based method of proofreading that breaks a work into individual pages to allow multiple proofreaders to work on the same book simultaneously. About half of the Project Gutenberg books have come out of this commons-based peer production.[76]

What if we organized similar "distributed transcribers" to work on handwritten historical documents that otherwise will never be digitized? Volunteers could take their turns transcribing page images of the widely used Cameron Family Papers at the Southern Historical Collection that would be presented to them online. The same automated checking process used by Ames Clickworkers or among distributed proofreaders could be applied. A similar approach could be taken to transcribing the massive quantities of recorded sound—the Lyndon B. Johnson tapes, for example—that are enormously expensive to transcribe and cannot be rendered into text with current automated methods. Max J. Evans, the head of the National Historical Publications and Records Commission, has recently proposed something similar. He called for a corps of "volunteer data extractors" who would index and describe archival collections that are currently only minimally processed. Such an approach, he argues, would take "advantage of organized, or self-selected and anonymous users who can work at home and in remote locations."[77]

The barriers to success in such a project are more social than technological. Devising the systems to present the page images or tapes online is not so difficult. It is harder to create the interest to involve volunteers in such a project. But who would have thought that 85,000 people would volunteer to look for Mars craters or that 60,000 people would write and edit entries for *Wikipedia*? Of course, denizens of the Internet are likely to be more excited about searching through Mars craters than through nineteenth-century women's diaries. Still, such projects have shown the ability, as Benkler wrote, to "capitalize on an enormous pool of underutilized intelligent human creativity and willingness to engage in intellectual effort."[78]

If the Internet and the notion of commons-based peer production provide intriguing opportunities for mobilizing volunteer historical enthusiasm to produce a massive digital archive, what about mobilizing and coordinating the work of professional historians in that fashion? That so much professional historical work already relies on volunteer labor—the peer review of journal articles, the staffing of conference program committees—suggests that professionals are willing to give up significant amounts of their time to advance the historical enterprise. But are they also willing to take the further step of abandoning individual credit and individual ownership of intellectual property as do *Wikipedia* authors?

Could we, for example, write a collaborative U.S. history textbook that would be free to all our students? After all, there is massive overlap in content and interpretation among the more than two dozen college survey textbooks. Yet the commercial publishing system mandates that every new survey text start from scratch. An open-source textbook would not only be free to everyone to read, it would also be free to everyone to write. An instructor dissatisfied with the textbook's version of the War of 1812 could simply rewrite those pages and offer them to others to incorporate. An instructor who felt that the book neglected the story of New Mexico in the nineteenth century could write a few paragraphs that others might decide to incorporate.

This model imagines something open and anarchistic in the style of *Wikipedia*. Textbooks (not to mention scholarly articles) pose deeper problems of mediating conflicting interpretation than are faced by *Wikipedia* with its factualist emphasis. But commons-based peer production need not be so unstructured. After all, not everyone can rewrite the Linux kernel core. Everyone can contribute ideas and codes, but a central committee decides what is incorporated in an official release. Similarly, *PlanetMath*, a free online collaborative math encyclopedia, uses an "owner-centric" authority model in contrast to *Wikipedia*'s "free form" approach. As one of the founders, Aaron Krowne, has explained, "there is an owner of each entry—initially the entry's creator. Other users may suggest changes to each entry, but only the owner can apply these changes. If the owner comes to trust individual users enough, he or she can grant these specific users 'edit' access to the entry." This has the potential disadvantage of discouraging open participation and requiring more commitment from some participants, but it gives a much stronger place to expertise by assuming that the "owner is the de facto expert in the topic at hand, above all others, and all others must defer."[79]

Even so, the difficulties in implementing such a model for professional scholarship are obvious. How would you deal with the interpretative disputes that are at the heart of scholarly historical writing? How would we allocate credit, which is so integral to professional culture? Could you get a promotion based on having "contributed to" a collaborative project? There are no easy solutions. But it is worth noting that contributors to open-source software projects are not motivated simply by altruism. Their reputations—and hence their attractiveness as employees—are often greatly enhanced by participation in such projects. And we do reward people for collaborative professional work such as service on an editorial board. Nor are collaborative projects as free and frictionless as their greatest enthusiasts like to maintain. There are significant organizational costs—what the economists call "transaction costs"—to creating and maintaining such projects. Someone has to pay for the servers and the bandwidth and install and update the software. *Wikipedia* would have never gotten off the ground without the support of Wales and Bomis. More recently, it has launched fund-raising campaigns to cover its substantial and growing expenses.

Still, *Wikipedia* and Linux show that there are alternative models to producing encyclopedias and software than the hierarchical, commercial model represented by Bill Gates and Microsoft. And whether or not historians consider alternative models for producing their own work, they should pay closer attention to their erstwhile competitors at *Wikipedia* than Microsoft devoted to worrying about an obscure free and open-source operating system called Linux.

Practicing History in New Media

Teaching, Researching, Presenting, Collecting

4

Historians and Hypertext

Is It More Than Hype?

WITH STEVE BRIER

"I N TEN YEARS OR SO," D. H. JONASSEN PREDICTED more than ten years ago in *The Technology of Text* (1982), "the book as we know it will be as obsolete as is movable type today." Jonassen is hardly the only techno-enthusiast to get carried away by the potential of electronic media to reshape the way we consume and read information. More than thirty years ago, Ted Nelson, who coined the term "hypertext," was already arguing that print books would be obsolete in just five years.

The 1980s did not see the withering away of the print book, though they did mark the emergence of a vast literature (most of it, ironically, in print form) on "hypertext." But the failure of hypertext and electronic books to live up to the promises and prophecies of techno-enthusiasts should not lead us to dismiss out of hand the possibilities of using computers and digital media to present the past. Even considering the existing technology, electronic history books do have a place on our shelves even if it is not yet time to toss our trusty paperbacks into the dustbin of history (books). In this essay, we want to report on our own effort at electronic publishing as a way of suggesting some of the advantages and a few of the limitations of this new medium.

Briefly summarized, our electronic history book, *Who Built America? From the Centennial Celebration of 1876 to the Great War of 1914*, published by

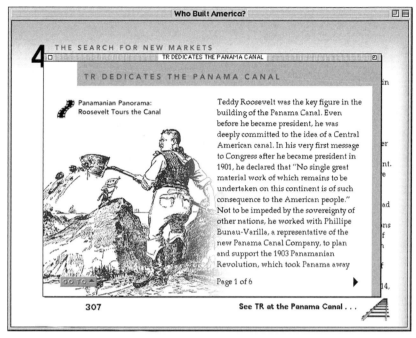

FIGURE 4.1 A sample screen from the CD-ROM version of *Who Built America?*. Clicking on the railroad track icon in the bottom right-hand corner takes users to a film clip depicting President Theodore Roosevelt during a trip to Panama.

the Voyager Company and developed by us in collaboration with Josh Brown and other colleagues at Hunter College's American Social History Project (ASHP) and George Mason University, provides an interactive, multimedia introduction to American history of the late nineteenth and early twentieth centuries on a single CD-ROM disk for Macintosh computers. The "spine" of this computer book is a basic survey of American history from 1876 to 1914, which is drawn from the second volume of ASHP's book, *Who Built America?* published by Pantheon Books in 1992. Added to—and, in the process, transforming—this textual survey are nearly two hundred "excursions," which branch off from the main body of the text. Those excursions contain about seven hundred source documents in various media that allow students as well as interested general readers to go beyond (and behind) the printed page and to immerse themselves in the primary and secondary sources that professional historians use to make sense of the past. In addition to about five thousand pages of textual documents, there are about four-and-a-half hours of audio documents (oral histories, recorded speeches, and musical performances), forty-five minutes of films, more than seven hundred photo-graphic-quality pictures, and about seventy-five charts, graphs, and maps.

The advantages of the computer and CD-ROM for presenting the turn of the century to students are easily summarized. One rests on the vast space offered by the CD-ROM, which can hold 640 megabytes of data—the equiv-alent of more than 300,000 typed pages. Thus, whereas the print version of four chapters of *Who Built America?* could only include 40 primary docu-ments of 250 to 750 words, the CD-ROM includes not only hundreds more documents but also much longer ones. For example, we have dozens of let-ters written home by Swedish, English, and Polish immigrants rather than just one or two. Indeed, the educational edition of *Who Built America?* in-cludes five full-length books, including Upton Sinclair's *The Jungle* and Booker T. Washington's *Up from Slavery*. (The capaciousness of the CD-ROM also encouraged us to incorporate some less serious excursions that might not make their way into a conventional book. One excursion explores the origin of the custom of answering the phone with "hello" and includes an early vaudeville routine, "Cohen on the Telephone." Another allows you to complete the world's first crossword puzzle, which appeared in a New York newspaper in 1913.)

The second key advantage of this medium is its ability to locate and keep track of all that information very efficiently. If we asked you to find the 117 instances of the word "work" in the first four chapters of the printed volume, it would take you a couple of days and you would probably miss some of them. In the electronic version, it would take only about 11 seconds. In just a few more seconds, we could even locate the 406 instances of the same word in all the thousands of pages of primary documents, the excursion introduc-tions, the captions, and the time line, and then provide a list of all the in-stances in context. The computer would also keep track of which ones you had looked at and keep markers for the ones to which you want to return. Moreover, you can search in more extensive and complex ways; you could be taken only to the instances of "working class" or "working-class politics" or even pages where the word "class" appears anywhere near the word "women." The program also offers many other ways to locate and link things very quickly. A "resource index," for example, provides you with rapid access to all 700 primary documents, whether organized by topic (from American Indi-ans to women) or by type of document (film, audio, photographs and im-ages, puzzles and games, maps and graphs, or text). Using the find feature and the resource index allows those who want to do so to learn about American history in a decidedly nonlinear fashion.

A related advantage of the computer's ability to access and keep track of information is what we might call "simultaneity"—or the ability to move very quickly from one body of information to another—to shift from reading the African American Congressman Robert Smalls denouncing the disfranchisement of black voters, to studying statistics on the effects of the secret ballot on voting, to examining the text of the Louisiana grandfather clause, to considering the historical debate between C. Vann Woodward and Howard Rabinowitz on the origins of Jim Crow, to listening to a recording of Booker T. Washington offering his famous "Atlanta Compromise." In effect, readers can instantly get behind the page to see materials out of which the basic text has been crafted. In addition, they can very quickly locate information that will help them understand what they are reading. You can, for example, hold down the mouse on any place name in the text and be shown that place's location on a U.S. map. If you are lost temporarily rather than spatially, you can switch to any of seven categories in a detailed time line that spans the 1876–1914 period.

The third thing that the electronic book can provide that the print version cannot is multimedia. For historians, the advantages of this are obvious. The past occurred in more than one medium, so why not be able to present it in multiple dimensions? The four-and-a-half hours of audio "documents" are particularly valuable in making the past immediate and vivid. Thirty-five different oral history witnesses—for example, Miriam Allen DeFord on her introduction to birth control in 1914; George Kills in Sight on the death of Crazy Horse; Pauline Newman on organizing the "Uprising of the 20,000" in 1909; Eubie Blake on the origins of ragtime; Vaseline Usher on the 1906 Atlanta race riot; and Luther Addington on religion in rural Appalachia—tell their stories in their own voices.

These reminiscences are supplemented by sixteen contemporary sound clips of famous Americans, including William Jennings Bryan delivering the "Cross of Gold" speech, Woodrow Wilson campaigning for the presidency, William Howard Taft appealing to labor voters, Andrew Carnegie touting the "Gospel of Wealth," Russell Conwell looking for "Acres of Diamonds," and Weber and Fields doing two of their popular vaudeville routines. Twenty-five songs present the musical diversity of the nation, including Mexican corridos, Sousa marches, Italian American Christmas music, Tin Pan Alley tunes, African American work songs, and coal miner laments.

When we first started work on this project in 1990, the only way to present film would have been to put it on an additional laserdisk to be played on a separate laserdisk player and television monitor. But the appearance in 1991 of "QuickTime," a piece of software that allows the display of films directly on the computer screen, has enabled us to incorporate twenty early film clips as well. Thus, the excursion on the 1912 election includes film footage of all three major party candidates. Other film footage includes Edwin Porter's "documentary" of the sinking of the *Maine*, women suffragists marching down Fifth Avenue in 1915, a 1904 view of the interior of the giant Westinghouse factory, black soldiers marching in the Philippines, heavyweight champ Jim Corbett knocking out Peter Courtney, and a streetcar view of Boston in 1906. We have also drawn upon the earliest fiction films: the excursion on the railroad allows you to see the film classic, *The Great Train Robbery*, in its entirety; and the excursion on temperance includes a satirical view of Carry Nation in *The Kansas Saloon Smashers*.

The advantages of massive space and multimedia bring with them some implicit dangers. There is no doubt, for example, that 640 megabytes offers exciting new possibilities that are simply not possible in 640 book pages. But as we know from print books, length does not equal quality. The old slogan from computer programming—"garbage in, garbage out"—can be equally applied to CD-ROMs. And while sound and film are terrific additions to history teaching, these media can turn history into television commercials in which the media glitz overwhelms sustained contact with difficult ideas. We believe that we should be very wary of the justification for multimedia that is offered over and over again: students no longer read; they are visually oriented. Therefore, we need to respond to that with multimedia. Well, yes, we should certainly respond, for example, by drawing on and then developing skills of visual literacy. But there is no reason to give up reading, simply because it is somehow old fashioned or out of fashion. Here, we heartily embrace the slogan that our collaborators at Voyager have coined and even put on their T-shirts: "Text: The Next Frontier."

Thus, despite the unconventional nature of *Who Built America?*, we have tried to retain some of the traditional features of a printed book. It even looks very much like a printed book on screen, with two columns of type and frequent pictures. We and our collaborators at Voyager devoted considerable time and effort to experimenting with typography and layout in order to make *Who Built America?* pleasing to look at and easy to read. The reader can page through the book briskly using the arrow keys. To make quick paging possible, the pictures are not immediately presented in full, eight-bit photographic quality, which takes a few seconds to load into the computer's memory. But the higher quality image, including a detailed caption and source information, is only a click of the mouse away.

Other traditional book features are also retained. You can, for example, take notes in the margin or in a separate notebook. (Less traditionally, a "resource collector" serves as a sort of multimedia notebook in which you can assemble your own compilation of specific sound and film clips and pictures as well as text.) If you prefer to do the electronic version of "highlighting," you can boldface or underline any text you select. Here the advantage over the printed book is that you can erase any markings you have made with no trouble. And you can electronically fold over ("dog-ear") the corner of a page and, again, the computer will rapidly locate that marked page for you as well as any of your marginal notes.

Just as *Who Built America?* on CD-ROM retains many of the traditional features of a print book, so too the process of writing it was quite traditional in many respects. The largest amount of our work involved the customary tasks that are familiar to all historians: selection, analysis, and synthesis. Whereas our CD-ROM contains many more primary documents than any print book, those documents still represent a small selection from the vast historical record. That selection is rooted in our best historical judgment. Similarly, the documents cannot readily stand on their own. Every document is explained and contextualized; we spent hundreds of hours researching the backgrounds for each of the documents and synthesizing the latest scholarship on a myriad of topics from sharecropping in the South, to Indian education at the Carlisle school, to Asian immigrants at Angel Island, to Coxey's Army in Washington, to the Armory Show in New York. We think there is little danger that the new electronic media are going to "displace" historians; a more realistic worry would be that the vast space of the CD-ROM is going to challenge the energy and ingenuity of historians to fill those disks in creative and intelligent ways.

New electronic media will also challenge our creativity as teachers if we are going to use them in ways that live up to their promise to democratize education and empower students. At its best, interactive technology has the potential to make exciting materials available to a broader audience, to give students and others direct access to primary documents that might be available at the Library of Congress but not at their local library. In addition, computer technology can make it possible (though it doesn't assure that possibility) for students and other readers to have more control over their learning and to move at their own pace and make decisions about what direction they want to go in, about what byways they want to investigate. New technology may also free up teachers from some of the most repetitive and least edifying aspects of teaching and allow them to spend time working directly and creatively with students.

But we need always to keep in mind the opposite tendencies that are at least implicit (e.g., expensive new technologies may wind up being available only at more affluent institutions, and the technologies can be exploited to make education more constraining, to make students move at an even more rigid lockstep through material). And at least to some, the new technology is seen as a cheap and quick fix to the problems of American education. Chris Whittle and his business and education allies are, for example, promoting technology as a way of increasing teacher productivity and (not coincidentally) turning education into a private and profit-making venture.

Our own view, however, is that we are a long way from the days when students will telecommute to Digital U. Even the seemingly simple and straightforward task of getting faculty and students to learn and make use of new computer technologies is far from simple or straightforward. And, at least for the next few years, our struggles with incompatible formats and platforms will remind us why the adoption of standard railroad track gauges was an important breakthrough of the late nineteenth century. Moreover, we doubt that teaching will be as readily subject to automation as some believe or hope. Still, the future is far from determined. Any new technology carries within it repressive as well as liberating possibilities. Although a Luddite resistance to technological change may seem appealing at times, we would argue instead that it is worth engaging with these new technologies in an effort to try to insure that they indeed become badly needed tools of empowerment, enlightenment, and excitement. We may still be a long way from the death of the print book, but the rapid emergence of new electronic media for reading, storing, and linking information means that even those of us who spend much of our time studying the past will need to pay close attention, as well, to the future.

5

Rewiring the History and Social Studies Classroom

Needs, Frameworks, Dangers, and Proposals

WITH RANDY BASS

W ITHIN FIVE YEARS OF ALEXANDER GRAHAM Bell's first display of his telephone at the 1876 Centennial Exposition, *Scientific American* promised that the new device would bring a greater "kinship of humanity" and "nothing less than a new organization of society." Others were less sanguine, worrying that telephones would spread germs through the wires, destroy local accents, and give authoritarian governments a listening box in the homes of their subjects. The Knights of Columbus fretted that phones might wreck home life, stop people from visiting friends, and create a nation of slugs who would not stir from their desks.[1]

Extravagant predictions of utopia or doom have accompanied most new communications technologies, and the same rhetoric of celebration and denunciation has enveloped the Internet. For *Wired* magazine publisher Louis Rossetto, the digital revolution promises "social changes so profound that their only parallel is probably the discovery of fire." According to Iraq's official government newspaper, *Al-Jumhuriya*, the Internet spells "the end of civilizations, cultures, interests, and ethics."[2]

The same excessive rhetoric has surrounded specific discussions of computers and education. "Thirty years from now the big university campuses will be relics," proclaims Peter Drucker in *Forbes*. "It took more than 200 years (1440 to the late 1600s) for the printed book to create the modern school. It won't take nearly that long for the [next] big change." One advertisement on the Web captures the mixture of opportunity and anxiety occasioned by the new technology. Three little red schoolhouses stand together in a field. A pulsing green line or wire lights up one of the schools with a pulse of energy and excitement, casting the others into shadow. "Intraschool is Coming to a District Near You," a sign flashes. "Don't Be Left Behind!" And the other side has similarly mobilized exaggerated forecasts of doom. Sven Birkerts, for example, laments new media as a dire threat to essential habits of wisdom—"the struggle for which has for millennia been central to the very idea of culture."[3]

There are some encouraging recent signs that the exaggerated prophecies of utopia or dystopia are fading and we are beginning the more sober process of assessing where computers, networks, digital media (our working definition of "technology") are and aren't useful. Rather than apocalyptic transformation, we seem to be heading toward what Phil Agre calls the "digestion model." "As new technology arises," he observes, "various organized groups of participants in an existing institutional field selectively appropriate the technology in order to do more of what they are already doing—assimilating new technology to old roles, old practices, and old ways of thinking. And yet once this appropriation takes place, the selective amplification of particular functions disrupts the equilibrium of the existing order, giving rise to dynamics internal to the institution and the eventual emergence of a new, perhaps qualitatively different equilibrium."[4]

In social studies education, we have already begun the process of "selective appropriation" of technology.[5] But before we can move to a new and hopefully better equilibrium, we need to ask some difficult questions. First, and most important: what are we trying to accomplish? Second, what approaches will work best? Third, are there dangers that we need to avoid as we selectively appropriate new technology for the social studies classroom? Fourth, how can we encourage and support the adoption and development of the best practices?

Why Use Technology in Social Studies Education?

Over the past five years of running technology workshops with hundreds, if not thousands, of college and pre-college teachers, we have usually begun by asking them: "What are you doing now in your teaching that you would like to do better? What do you wish your students did more often or differently?" "What pedagogical problems are you looking to solve?" Most commonly, they say they want their students more engaged with learning; they want students to construct new and better relationships to knowledge, not just represent it on tests; and they want students to acquire deeper, more lasting understanding of essential concepts.

Such responses run counter to another public discourse about history and social studies education—the worry, if not alarm, about student knowledge of a body of factual material. "Surely a grade of 33 in 100 on the simplest and most obvious facts of American history is not a record in which any high school can take pride," goes a lament that anyone who follows social studies education will find familiar. Indeed, it should be familiar: this particular quote comes from a study published in the *Journal of Educational Psychology* in 1917. As educational psychologist Sam Wineburg points out, "considering the differences between the elite stratum of society attending high school in 1917 and the near universal enrollments of today, the stability of this ignorance inspires incredulity. Nearly everything has changed between 1917 and today except for one thing: kids don't know any history."[6] Also unchanged is the persistent worry by school boards and public officials about that seeming ignorance.

And yet based on our own experience, this is not the problem that most concerns those teaching in our classrooms (except insofar as curriculum standards and exams constrain innovation and flexibility); neither is it the problem that most concerns those who have studied in those classrooms. In 1994, we undertook a nationwide study of a representative cross-section of 808 Americans (as well as additional special samples of 600 African Americans, Mexican Americans, and Sioux Indians) that sought to uncover how Americans use and understand the past. We asked a portion of our sample "to pick one word or phrase to describe your experiences with history classes in elementary or high school." Negative descriptions significantly outweighed positive ones. "Boring" was the single most common word offered. In the entire study, the words "boring" or "boredom" almost never appeared in descriptions of activities connected with the pursuit of the past, with the significant exception of when respondents talked about studying history in school—where it comes up repeatedly.[7]

The same point came across even more sharply when we asked respondents to identify how connected with the past they felt in seven different situations—gathering with their families, celebrating holidays, reading books, watching films, visiting museums or historic sites, and studying history in school. Respondents ranked classrooms dead last with an average score of 5.7 on a 10-point scale (as compared, say, with 7.9 when they gathered with their families). Whereas one-fifth of respondents reported feeling very connected with the past in school (by giving those experiences a rank of 8 or higher), more than two-thirds felt very connected with the past when they gathered with their families. Of course, the comparison we posed is not an entirely fair one. Schools are the one compulsory activity that we asked about; the others are largely voluntary (though some might disagree about family gatherings). Still, our survey finds people most detached from the past in the place that they most systematically encountered it—the schools.

To be sure, these negative comments about classroom-based history were not always reflected in remarks about specific teachers. Respondents, for example, applauded teachers for engaging students in the study of the past through active learning. A North Carolina man in his mid-twenties, for example, praised a teacher who "got us very involved" because she "took us on various trips and we got hands-on" history. A Bronx woman similarly talked enthusiastically about the "realism" of a class project's engagement with an incident in Puerto Rican history: "Everybody had different information about it, and everyone was giving different things about the same thing, so it made it very exciting."

Although teachers could make history classrooms resemble the settings in which, and the ways that, respondents liked to engage the past, most Americans reported that history classrooms more often seemed to include a content that was removed from their interests and to feature memorization and regurgitation of senseless details. Respondents recalled with great vehemence how teachers had required them to memorize and regurgitate names, dates, and details that had no connection to them. They often added that they forgot the details as soon as the exam had ended. Such complaints could be captured in the words of a 36-year-old financial analyst from Palo Alto, California: "It was just a giant data dump that we were supposed to memorize . . . just numbers and names and to this day I still can't remember them."

Not everyone would agree with these complaints. Others would argue that the real problem of the schools is historical and civic illiteracy—a lack of knowledge of the basic facts about history, politics, and society. Our own view (and that of the teachers with whom we have worked) is that such factual knowledge emerges out of active engagement with learning rather than out of a textbook and test-driven curriculum. Given that these are contentious issues, we think that it is important to acknowledge our bias up front. The problem we seek to address is the one that preoccupies the teachers with whom we have worked and the survey respondents with whom we talked—how can the social studies classroom become a site of active learning and critical thinking? Can technology foster those goals?

What Works? Three Frameworks for Using Technology to Promote Active Learning

The encouraging, albeit anecdotal, news from the field is that technology has, in fact, served those goals for a number of teachers and students across the country and that there is an emerging body of experience that suggests some of the most promising approaches. Our own framework for categorizing and discussing these approaches grows out of our observation of scores of teachers in workshops sponsored by the American Studies Crossroads Project, the New Media Classroom, and the Library of Congress's American Memory Fellows program.[8] Based on these interactions, we have concluded that the most successful educational uses of digital technology fall into three broad categories:

> *Inquiry-based learning* utilizing primary sources available on CD-ROMs and the World Wide Web, and including the exploration of multimedia environments with potentially fluid combinations of text, image, sound, moving images in presentational and inquiry activities;
>
> *Bridging reading and writing through online interaction*, extending the time and space for dialogue and learning, and joining literacy with disciplinary and interdisciplinary inquiry;
>
> *Making student work public in new media formats*, encouraging constructivist pedagogies through the creation and exchange of knowledge representations, and creating opportunities for review by broader professional and public audiences.

Each type of activity takes advantage of particular qualities of the new media. And each type of activity is also linked to particular pedagogical strategies and goals.

Inquiry Activities: The Novice in the Archive

Probably the most important influence of the availability of digital materials and computer networks has been on the development of inquiry-based exercises rooted in the retrieval and analysis of primary social and cultural documents. These range from simple Web exercises in which students must find a photo that tells something about work in the late nineteenth century to elaborate assignments in which students carefully consider how different photographers, artists, and writers historically have treated the subject of poverty. Indeed, teachers report that inquiry activities with digital materials have been effective at all levels of the K–12 curriculum. In Hillsborough, California, for example, middle school students simulate the work of historians by closely analyzing images of children at the turn of the century that can be found online. They then build from that to a semester-long project that asks students to "construct an understanding of the major 'themes' of the period and how these might impact a child born in 1900." To do that they must assemble a physical and digital scrapbook of letters, images, oral histories, artifacts, and diary entries and think critically about those sources.[9] Similarly, fourth graders in New York use the Works Progress Administration life histories online at the Library of Congress to reconstruct the worlds of immigrants and then use photographs from online archives to "illustrate" these narratives in poster presentations. And high school juniors in Kansas City scrutinize the "Registers of Free Blacks" at the Valley of the Shadow Civil War Web site not only to learn about the lives of free African Americans in the Shenandoah Valley before the Civil War but also to reflect on the uses and limitations of different kinds of digital and primary materials to achieve an understanding of the past.[10]

The analysis of primary sources, and the structured inquiry learning process that is often used in such examinations, are widely recognized as essential steps in building student interest in history and culture and in helping them understand the ways that scholars engage in research, study, and interpretation. Primary documents give students a sense of the reality and the complexity of the past; they represent an opportunity to go beyond the pre-digested, seamless quality of most textbooks to engage with real people and problems. The fragmentary and contradictory nature of primary sources can be challenging and frustrating, but also intriguing and ultimately rewarding, helping students understand the problematic nature of evidence and the constructed quality of historical and social interpretations. Virtually all versions of the national standards for history and social studies published in the 1990s have (in this regard, at least) followed the lead of the 1994 *National Standards for United States History*, which declared that "perhaps no aspect of historical thinking is as exciting to students or as productive of their growth as historical thinkers as 'doing history,'" by directly encountering "historical documents, eyewitness accounts, letters, diaries, artifacts [and] photos."[11]

Of course, the use of primary sources and inquiry methods does not require digital tools. Teachers have long used documentary anthologies and sourcebooks (often taking advantage of another somewhat less recent technological advance, the Xerox machine). But the rise of new media and new computer technology has fostered and improved inquiry-based teaching for three key reasons.

First and most obviously is the greatly enhanced access to primary sources that CD-ROMs and the Internet have made possible. Almost overnight teachers, school librarians, and students who previously had scant access to the primary materials from which scholars construct interpretations of society and culture now have at their disposal vast depositories of primary cultural and historical materials. A single Internet connection gives teachers at inner-city urban schools access to more primary source materials than the best-funded private or suburban high school in the United States. Just the sixty different collections (containing about one million different primary documents) that the Library of Congress has made available since the mid-1990s constitute a revolution in the resources available to those who teach about American history, society, or culture. And almost weekly, major additional archives are coming online. These include such diverse collections as the U.S. Supreme Court Multimedia Databases at Northwestern University (with its massive archive of written and audio decisions and arguments before the Court), the U.S. Holocaust Memorial Museum (with its searchable database of 50,000 images), and George Mason University's Exploring the French Revolution (with its comprehensive archive of images and documents).[12]

For the history and social studies teacher and the school librarian, even the most frequently criticized feature of the Web—the unfiltered presence of large amounts of junk—is potentially an opportunity, albeit one that must be approached with care. Bad and biased Web sites in the hands of the creative teacher are fascinating and revealing primary sources. In effect, many skills traditionally taught by social studies teachers—for example, the critical evaluation of sources—have become even more important in the online world. The Web offers an exciting and authentic arena in which students can learn to become critical consumers of information. Equally important, the Web presents the student with social knowledge employed in a "real" context. A student studying Marcus Garvey or Franklin Roosevelt through Web-based sources learns not simply about what Garvey or Roosevelt did in the 1920s and 1930s but also what these historical figures mean to people in the present.

A second appealing feature of this new distributed cultural archive is its multimedia character. The teacher with the Xerox machine is limited to written texts and static (and perhaps poorly copied) images. Now, teachers can engage their students with analyzing the hundreds of early motion pictures placed online by the Library of Congress, the speeches and oral histories available at the National Gallery of Recorded Sound that Michigan State is beginning to assemble, and hundreds of thousands of historical photographs.[13]

Third, the digitization of documents allows students to examine them with supple electronic tools, conducting searches that facilitate and transform the inquiry process. For example, the American Memory Collection provides search engines that operate within and across collections; if one is researching sharecropping in the thousands of interview transcripts held in the Federal Writers' Project archive, a search can quickly find (and take you to) every mention of sharecropping in every transcript. Similarly, searches for key words such as "race" or "ethnicity" turn up interesting patterns and unexpected insights into the language and assumptions of the day. In other words, the search engines cannot only help students to find what they are looking for; they also allow them to examine patterns of word usage and language formation within and across documents.

These kinds of activities—searching, examining patterns, discovering connections among artifacts—are all germane to the authentic thinking processes of historians and scholars of society and culture. Digital media not only give flexible access to these resources but also make visible the often-invisible archival contexts from which interpretive meaning gets made. "Everyone knows the past was wonderfully complex," notes historian Ed Ayers. "In conventional practice, historians obscure choices and compromises as we winnow evidence through finer and finer grids of note-taking, narrative, and analysis, as the abstracted patterns take on a fixity of their own. A digital archive, on the other hand, reminds us every time we look at it of the connections we are not making, of the complications of the past."[14]

The combination of increased access with the development of powerful digital searching tools has the potential to transform the nature and the scale of students' relationship to the material itself. For the first time, perhaps, it allows the novice learner to get into the archives and engage in the kinds of archival activities that only expert learners used to be able to do.[15] Of course, the nature of their encounter with primary materials and primary processes is still as novice learners. The unique opportunity with electronic, simulated archives is to create open but guided experiences for students that would be difficult or impractical to re-create in most research library environments. It also frees students and teachers from the traditional dependence on place for firsthand social, political, or historical research. Or, perhaps more importantly, it means that students can more readily compare their own community with others, more distant.

The task of creating these open but guided experiences is demanding. Teachers must not only learn how to use the new technology but also spend time exploring the digital archives (perhaps in partnership with school librarians) in order to learn what they hold and consider what students can learn from them. The construction of effective inquiry activities demands knowledge of the topic, the documents, the archive, and the craft of introducing students to the inquiry process. Implementing inquiry approaches in the classroom takes considerable class time, which teachers are sometimes reluctant to give. And the inquiry process is by definition not easy to control; students are likely to come up with unanticipated answers. At their best, however, new media technologies can help make the "intermediate processes of historical cognition" visible and accessible to learners, in part by helping students approach problem-solving and knowledge-making as open, revisable processes and in part by providing tools to give teachers—as expert learners—a window into student thinking processes.[16]

Bridging Reading and Writing Through Online Interaction

One very significant dimension of "making thinking visible" is the bridging of reading and writing through online writing and electronic dialogue. The benefits of writing and dialogue for student learning were well established before the emergence of computers and the Internet. Over the last several decades, educators in many disciplines and at every level of education have come to believe that meaningful education involves students not merely as passive recipients of knowledge dispensed by the instructor but as active contributors to the learning process. One of the key elements in this pedagogy is the importance of student discussion and interaction with the instructor and with each other, which provides opportunities for students to articulate, exchange, and deepen their learning. Educators in a wide range of settings practice variations of this process.

But the emergence of digital media, tools, and networks has multiplied the possibilities. Electronic mail, electronic discussion lists, and Web bulletin boards can support and enhance such pedagogies by creating new spaces for group conversations.[17] One of the greatest advantages to using electronic interaction involves the writing process, which can facilitate complex thinking and learning as well as build related skills. These advantages can combine with the potential for electronic discussion to draw out students who remain silent in face-to-face discussion. Online interaction has also proven to be effective in helping to build connections between subject-based learning and literacy skills (reading and writing), which too often are treated separately.

Online discussion tools also foster community and dialogue. Active, guided dialogue helps involve students in the processes of making knowledge, testing and rehearsing interpretations, and communicating their ideas to others in public ways. Online dialogue tools have yet another advantage in helping students make connections beyond the classroom, whether by enhancing the study of regional and national history through connections with a classroom elsewhere in the United States or by helping to expand global social studies curricula through e-mail "penpal" programs with students elsewhere in the world. Postcard Geography is a simple project, organized through the Internet, in which hundreds of classes (particularly elementary school classes) learn geography by exchanging postcards (real and virtual, purchased and computer generated) with each other. An Alabama elementary school teacher notes the galvanizing effect of the project on her rural students who "don't get out of their city, let alone their state or country!"[18] At North Hagerstown High School in Maryland, high school students mount online discussions of issues such as the crisis in Kosovo, engaging in dialogue among themselves and with more far-flung contributors—from Brooklyn to Belgrade.[19]

Designing Constructive Public Spaces for Learning

Closely connected to both online writing and inquiry activities is the third dimension of our framework: the use of constructive virtual spaces as environments for students to synthesize their reading and writing through public products. As with the other uses of new technology, the advantages of public presentations of student work are well known. But, here again, the new technology—in particular the emergence of the Web as a public space that is accessible to all—has greatly leveraged an existing practice. Virtual environments offer many layers of public space that help make thinking visible and lead students to develop a stronger sense of public accountability for their ideas. The creation of public, constructed projects is another manifestation of these public pedagogies, one that engages students significantly in the design and building of knowledge products as a critical part of the learning process.

In the use of new media technologies in culture and history fields, "constructivist" and "constructionist" approaches provide ways for students to make their work public in new media spaces as part of the learning process, ranging from the individual construction of Web pages to participation in large, ongoing collaborative resource projects that involve many students and faculty over many years of development.[20] For example, at an elementary school in Virginia, fifth graders studying world cultures build a different "wing" of a virtual museum each year, researching and annotating cultural artifacts and then mounting them online. Similarly, at a middle school in Philadelphia sixth graders worked closely with a local museum to create a CD-ROM exhibit on Mesopotamia, using images and resources from the museum's collections.[21] Seventh graders in Arlington, Virginia, published an online "Civil War Newspaper" with Matthew Brady photographs from the Library of Congress together with their own analyses of the photos.[22] More ambitious student-constructed projects can evolve over several years and connect students more closely to their communities, as in St. Ignatius, Montana, where high school students have helped to create an online community archive.[23]

The power of the digital environment for these kinds of projects comes not merely from their public nature but also from the capabilities of electronic tools for representing knowledge in nonlinear ways and through multiple media and multiple voices. Digital tools can represent complex connections and relationships and make large amounts of information available and manipulable. We have only begun to understand how to use digital tools for constructionist learning approaches that help students acquire and express the complexity of culture and history knowledge. Student constructionist projects offer a potentially very rich synthesis of resources and expressive capabilities; they combine archival and database resources with conversational, collaborative, and dialogic tools, in digital contexts characterized by hypertext and other modes for discovering and representing relationships among knowledge objects.

What to Avoid: Hazards Along the Electronic Frontiers

These are all appealing goals and there is some encouraging, although still preliminary, experience to suggest that technology can help us achieve them. But it would be foolish, if not dangerous, to suggest that technology is either a panacea for the problems of history and social studies education or that any of these approaches is easy to implement. Indeed, the most serious danger from the introduction of technology into the classroom is the mistaken assumption that it, alone, can transform education. The single-minded application of technological solutions to teaching will surely be as much of a disaster as the application of single-minded solutions to agriculture or forest management. As the first generation of scientific foresters learned, any change in a complex environment needs to be thought out ecologically.[24] New technologically enhanced approaches—whether inquiry-based learning or student constructionist exercises—must be carefully introduced within the context of existing teaching approaches and existing courses and assignments. What assignments are already working well? How will a new assignment alter the overall balance of a course? How do new approaches manifest themselves throughout a curriculum or a school?

In asking these questions, we should also be reminding ourselves to use technology only when it makes a clear contribution to classroom learning. Some teaching strategies, for example, work better with traditional materials. A teacher who has his students post rules of historical significance on butcher paper around the classroom may find that their visual presence is stronger on the classroom walls than on the class Web site. More generally, technology is better employed to provide a deeper understanding of some pivotal issues through inquiry and constructionist assignments rather than being pressed into service to respond to standards-based pressures for coverage.

By always thinking about whether new technologies respond to the goals with which we began, we can also be alert to the situations where technology might operate in the opposite direction from the one intended. It is important to acknowledge that while there are plenty of positive experiences with technology on which to draw, there is also a large body of negative examples from which we also need to learn. The most obvious examples can be found in the considerable volume of educational software that promotes passivity rather than the much-promised "interactivity." One of the great advantages of digital media—the ability to incorporate sound and film with text and images—is also one of the greatest problems because of the temptation to turn history into TV commercials in which the media glitz overwhelms sustained contact with difficult ideas. This has been the case with some multimillion-dollar multimedia extravaganzas that offer multiple interpretations of topics without giving the user any sense of which interpretations are more plausible than others or without any real level of interactivity that encourages active and critical thinking.

Some of these same tendencies were also embodied in the worst of the CD-ROMs that appeared on the market in the early and mid 1990s. In many, the notion of multimedia was a voice reading words that already appeared on the screen. One CD-ROM (which sold for $395) turned out to be a recycled filmstrip—and a twenty-five-year-old one at that.[25] Such uses of digital media promote the same deadening memorization of facts that generations of students have complained about and waste scarce school funds on the products of sleazy educational hucksters.

The pressure of commercial vendors leads to a related pitfall: the possibility that school systems will invest in equipment, software, and narrowly defined technological training at the expense of funding professional development to use new technology wisely. Computers are expensive, delicate machines that break down often and require recurrent maintenance. The rapid development of the field means that computer labs quickly become outdated. Wiring classrooms for Internet access is expensive and sometimes difficult, particularly in older school buildings. Software can also be costly, and the constant updates required to stay in step with new resources highlight the need for instructional technology staff. Providing effective staff development for teachers throughout the educational system would add significantly to the cost of purchasing hardware. The combined expense of installing, maintaining, and supporting the effective use of operative computer labs can be overwhelming. And, as Diane Ravitch rightly points out, "the billions spent on technology represent money not spent on music, art, libraries, maintenance and other essential functions."[26]

Such costs weigh unevenly on different schools, school systems, and communities—another key threat that new technology poses. Under-resourced schools and colleges have a particularly difficult time finding the funds to pay the price required for new technology. Although federal, state, and corporate grant programs are helpful, they are not sufficient, and they usually pay only for hardware, not for maintenance or staff development. As a result, the schools and colleges serving poor and working-class communities lag behind in the effective implementation of technology. And their students—disproportionately African American or Latino—are the ones that suffer most. According to the most recent report from the National Center for Education Statistics, 51 percent of public school classrooms nationwide have Internet access. But for schools with large numbers of poor or minority students, the number drops to less than 40 percent. This disparity shapes colleges and universities as well. Whereas 80.1 percent of all students entering elite private colleges report they use computers regularly, only 41.1 percent of students entering historically black colleges report similar usage. In many colleges, students who come from under-resourced school systems will find technology to be one more item to be added to an already daunting list of educational and social challenges. There is a real—and in many ways a growing—threat that new technology will add to the already immense nationwide stratification of educational opportunity. Indeed, the most recent national report on the digital divide indicates that technology use continues to split along lines of both class and race.[27] And the problem is even worse when considered internationally.

Finally, there is the larger danger that educators, parents, and school boards come to see technology as an end in itself rather than a means to achieving better student learning. Technology can act as a powerful narcotic that lulls us into believing that we are teaching students to think simply by putting machines into classrooms. The hardest intellectual and pedagogical problems—teaching students to judge the quality of information, to deal with conflicting evidence, to develop analytical frameworks—are present in both the print and digital environments.

What Next? Toward Student Learning

Not surprisingly, our recommendations for the future grow out of our experience with this decade-long history of digital technology in the history and social studies classroom.

First, we would urge a renewed national commitment to insuring that the benefits of new technology be shared equally. Many others have made the same point, and there is little need to belabor it here.

Second (and while we are still tilting at windmills), we would argue that assessment needs to be revised to measure learning accurately in the new media environment. Right now, standards and assessment tend to hinder the integration of technology into teaching. When assessment, as in most states, requires pre-twentieth century technology (i.e., pen and paper) and is focused on content and factual knowledge, teachers are understandably reluctant to adopt strategies that take advantage of the potential of technology to promote deeper understanding. But if the assessment were designed to reflect deeper understanding of reading, interpreting, and arguing processes as well as what students need to know in the twenty-first century—including how to use the Internet and computers to research, analyze, and present information—then the integration of technology into the social studies and other academic curricula would be greatly fostered.

Third, we think that teachers need more tools and supports that will enable them to use electronic resources actively and critically. Teachers value gateway sites because they provide reliable starting points, filtering mechanisms, and sample curricula for using the Web.[28] In addition, since many teachers are themselves relative novices in the archives, they need guides to evaluating and analyzing primary source materials. They also need the kinds of software tools that allow their students to collaborate electronically with ease. And they need access to software and hardware that makes student constructionist projects feasible in multiple settings. Such software environments need to remain open and flexible, and not "one-size-fits-all" templates that presuppose certain teaching styles or approaches.

Fourth, teachers need robust professional development programs that will allow them to retool for the electronic future. The billions of dollars invested in preparing schools for the twenty-first century have gone (and continue to go) overwhelmingly to hardware and wiring. Where teachers lack necessary training and support, computer labs frequently wind up gathering dust, or being used as glorified typing labs. If meaningful progress is to occur in this field, funding for professional development must be given equal priority with funding for hardware. But it is not simply a matter of the quantity of available faculty development; it is also a question of quality. Typically, professional development in technology focuses narrowly on building skills or familiarizing teachers with particular software applications. The most common faculty development structure is a two- to four-hour workshop led by technology support staff who are skilled in technical issues but distant from the latest thinking about disciplinary content and teaching methodology. Our experience and feedback from our colleagues suggest the importance of developing a different approach.

In particular, we would encourage leaders in the field to create, nurture, and support professional development approaches that are rooted in the issues and experiences of everyday classroom practice, build directly on teachers' expertise in nontechnological settings, and enable teachers to adapt their skills to a new context. They need to speak to real classroom needs, helping teachers find ways to use technology to solve long-standing problems, do their work better, and more effectively reach their goals for their courses and their students. And they must point teachers towards classroom implementation, testing, and experimentation with real students in real classroom situations. In addition, professional development must involve a sustained and recursive process. Instead of one-shot workshops, effective professional development with technology must unfold over time and provide multiple opportunities for teachers to move back and forth between initial training workshops, classroom testing, and reflective seminars where they can articulate and collectively analyze their experiments using new technology resources.

Such approaches will themselves benefit from the effective uses of technology. One of the most exciting effects of the Internet for teachers has been the erosion of the isolation that traditionally afflicts the classroom teacher. The teachers with whom we have worked in Crossroads, the New Media Classroom, and the American Memory fellows program have acquired a much broader set of colleagues than was possible before. On a regular basis they consult with each other on how to teach a particular subject or to organize a particular assignment. Other teachers have developed mutually supportive relationships with teachers across the country they have never met but with whom they converse through lists like H-High, H-Teach, the "Talking History" forums sponsored by History Matters, or "Highroads," sponsored by Crossroads. In some of these settings, the high school teacher in Kansas City can get advice on the latest developments in women's history from a leading scholar such as Gerda Lerner or find out about successful assignments from an award-winning high school teacher from Virginia. The often chaotic information environment of the Web also encourages teachers to forge partnerships with school librarians, who can bring particular skills in information evaluation to the table.

Fifth, given the difficulty of altering entrenched patterns of professional development, it makes sense to focus efforts on pre-service education as well as in-service. Such efforts—as manifest in education curricula and state certification requirements—need to go considerably beyond courses on new media and teaching methods. Future teachers most need discipline-based courses in which technology is integrated into the course content. Such courses can enable teachers to understand the archive-at-a-mouse-click not as some new way to bring the library to the doorstep, but as a fundamental shift in how society handles knowledge—its accessibility and what one can do with it. Moreover, teachers will never make effective use of the vast archives now accessible to them unless they understand, for example, the nature of historical evidence and argumentation or other disciplinary contexts for using new media.[29] More generally, the educating of teachers to use technology effectively must go far beyond simple training in software or techniques for implementation to include an initiation into habits of reflective practice that will allow them to adapt and innovate in new learning environments throughout their careers, even as specific technologies and applications change.

Sixth, we need to acknowledge that we are still at the starting point of the selective appropriation of new technology and that we need serious classroom research into what does and doesn't work. Some of this research needs to come from professional educational researchers. But we also believe that research can be combined with professional development where the teacher becomes the researcher. Supportive of this recommendation are approaches that have begun to emerge on the college level under the rubric of the "scholarship of teaching" and are beginning to be explored on the pre-college level as well.

But whatever approaches are taken, we need to return continually to first principles and ask ourselves: what are we trying to accomplish in the history and social studies classroom? Can technology help to make that possible? One way to keep that mantra in mind is to recall the old joke about a man who works in a factory and leaves there every evening with a wheelbarrow full of straw. Every night as he exits the factory and passes through the gate, the guard looks through the straw, certain that the man is stealing something. At the end of twenty years' employment, the man is departing, as always with his wheelbarrow full of straw. The guard turns to the man and says:

"For twenty years you have been leaving every night with a wheelbarrow full of straw. For twenty years, every night, I look through the straw and find nothing. I know you have been stealing something. This is your last night. For my own curiosity, you have to tell me: what have you been stealing all these years?"

The man replies, "Wheelbarrows."

If that joke were taken as an analogy, then technology is the straw. It is merely the prop by which we are getting something more valuable (the wheelbarrow) out the door. And what are the more valuable things we're trying to get out the door? They are, we would argue, the enhancement of learning through interaction and dialogue; an increasingly expansive, in- clusive, and socially conscientious approach to the study of history, society, and culture; and the elevation of our standards for what passes as student learning.

6

The Riches of Hypertext for Scholarly Journals

C HANGE COMES SLOWLY IN ACADEMIC LIFE.
Place the *American Historical Review* from 1899 next to the current, late-1999 issue, and you'll discover a reassuring continuity. To be sure, the content of the articles has been transformed, with "Creole Bodies in Colonial South Africa" displacing "Connecticut Loyalists." The selections have also gotten a bit longer, but not drastically so. But the form of the journal and the articles remain largely unchanged—monographic articles are at the front of the journal, book reviews at the back; footnotes at the bottom of the page; about 500 words to the page; articles put their thesis statement at the front, conclusions at the end. If the journal's J. Franklin Jameson were to rise from the dead (I'm told that someone once called the American Historical Association to report that she was channeling the former editor), he would be able to resume his duties without too much trouble.

Conventional wisdom tells us that computers and the Internet will rapidly change all that. But just how will scholars and, in particular, scholarly journals "digest" the new technology? That is what colleagues and I from the Center for History and New Media at George Mason University and two leading humanities journals—the *Journal of American History* and *American Quarterly*—have been wrestling with over the past year or two. In a Mickey

Rooneyesque spirit of "Let's put on a show," we decided to try out some things in electronic publishing, rather than simply sponsor more theorizing about what the cyberfuture might look like. I'd like to look back on what we learned, as a way of looking forward to where we go next.

Our efforts, of course, built on considerable (and better-financed) experimentation that has already taken place. Not surprisingly, the most rapid changes have come in journals of science and technology, where it looks as though electronic-only publication will become increasingly common. Here, the model of instant publication pioneered by Paul H. Ginsparg, a physicist at the Los Alamos National Laboratory, whose database of scientific papers is growing at the rate of 25,000 per year, has been particularly influential. Last year, for example, the National Institutes of Health announced plans for PubMed Central, a similar online archive of scholarly papers in the life sciences.

Cyberjournals—covering such diverse topics as textual criticism of the Bible and postmodern culture—have emerged in the humanities as well. But print humanities journals are likely to survive, in part because of the importance of narrative or discursive articles (as opposed to reports on research findings), which are often consumed at a leisurely pace, away from the computer screen. Nor do humanists care about scholarly currency in quite the same way that physicists or doctors do. Last year's reflection on Jane Austen remains a good deal more compelling to humanists than last year's study of adverse drug reactions is to scientists. At least until breakthroughs in screen display make onscreen reading easier, the most important experiments are likely to be hybrids in which digital publication supplements print versions.

Before long, it seems likely, every print journal will have its electronic clone. Indeed, such clones already exist, even for many journals that have not created them explicitly. Commercial operations like Bell & Howell's ProQuest and Northern Light offer electronic versions of articles, on a per-article basis, from hundreds of scholarly journals—thereby "unbundling" the carefully bound products that journal editors have crafted. Ostensibly, no electronic edition of the *Journal of Modern History* exists, but you can, in fact, read it online through ProQuest.

Such changes, however, do not alter the essential intellectual product offered by the journals. Do electronic media allow us to do anything different from what journals have done for the past century?

One obvious opportunity is for what I would call the "digital supplement," in which we take advantage of the cheapness of digital storage to make available materials that are of interest to specialized audiences, but cannot be provided economically in print. Most scholarly work involves the creation of an "archive" of some sort, although generally that archive remains stowed away in an individual scholar's file cabinet or computer.

The *Journal of American History*'s March 1999 roundtable on interpreting the Declaration of Independence through translation was a natural candidate for exploring the possibilities of the digital supplement. Although the print journal was able to devote a substantial number of pages to the roundtable, it could not also include the many versions of the Declaration of Independence, translated into different languages at different times, that the journal's authors had assembled in the process of writing their articles. On the World Wide Web (http://chnm.gm.edu/declaration), we were able to include that richer documentation. Where possible, we also included retranslations back into English, so that readers who didn't know the various languages could get a sense of how some key concepts and words had been rendered in translation.

Two other features of the project also recommended it for online publication. First, we wanted it to be an open-ended and evolving effort, and welcome contributions of other translations, along with commentary about them. Versions of the Declaration of Independence in Bulgarian, Turkish, and Estonian now seem to be in the offing. Second, given the international character of the project, it seemed particularly appropriate to use the Web, a medium that has allowed us to attract significant readership over the past year from outside the United States.

We have also gotten an unanticipated—but quite significant—result: a broader audience for a journal that has traditionally reached professional historians. While nonprofessionals surely wander across print copies of the journal in large public libraries, it seems, anecdotally, that our open-publication format has attracted a greater share of nonacademic readers. Indeed, all of the e-mail I have received about the issue has come from nonhistorians. A high-school English teacher from Texas, for example, wrote to say that he was going to use the translations in a class exercise on the Declaration as literature.

Such projects easily establish the way that we can do more online—offer fuller documentation, reach larger audiences. But can we do anything different? That is the distinction drawn by the scholar Janet H. Murray, in her book *Hamlet on the Holodeck: The Future of Narrative in Cyberspace*, between "additive" and "expressive" features of new media. She makes the useful analogy to early films, which were initially called "photoplays" and were thought of as additions that combined photography with theater. Only when film-makers learned to use such techniques as montage, close-ups, and zooms as a part of storytelling did photoplays give way to the new, expressive form of movies.

What would a scholarship that made expressive rather than just additive use of new media look like? That was the goal of a more ambitious project that we undertook with *American Quarterly* (http://chnm.gmu.edu/aq). In setting up the project, we wanted to encourage unconventional departures in form, while retaining the conventional validation and peer review that characterize scholarly publication. But we soon realized that it would be unfair to ask people to submit hypertext essays on speculation. If, as would normally be the case, we rejected many of them, the authors would have almost no other journals to which they could send their work. Our compromise was to invite people to submit proposals for online scholarly articles. Out of twenty proposals, we gave a go-ahead to four.

The authors took differing approaches to putting scholarship online. Thomas Thurston's "Hearsay of the Sun: Photography, Identity, and the Law of Evidence in Nineteenth-Century American Courts" does something that, at first glance, seems straightforward—but is rarely, if ever, accomplished in print scholarship. It presents not just an argument about the legal status of photography in the nineteenth century but also virtually all of the evidence that underlies (or even undercuts) the argument. One can read Thurston *and* one can read forty-two court decisions, articles, and excerpts from various novels and legal treatises. Of course, given enough pages (not a small matter), that might be done as well in a print journal.

But Thurston also offers us something else: a system for seamlessly linking argument and evidence, a new scholarly technology, if you will. He does that by taking advantage of two simple features of Web browsers: the "anchor" tag, which makes it possible to move the reader directly from one reference to the paragraph from which it originated, and the "frame," which enables Thurston to keep all of the pieces (argument, footnotes, sources, illustrations) of his article on a single screen.

James Castonguay's "The Spanish-American War in U.S. Media Culture" also provides us with a scholarly innovation in the way that it connects evidence and argument. The "illustrations" include actual films rather than just film stills.

For a scholar of comics, as well, the Web offers the opportunity to transcend the limits of print publication. A print journal would find it prohibitively expensive to include the more than fifty color illustrations that accompany David Westbrook's "From Hogan's Alley to Coconino County." Moreover, Westbrook enables a kind of simple interactivity that print cannot easily replicate—he encourages the reader to interact with the evidence and test his or her ability to see what the seasoned scholar notices.

Take a look at Richard F. Outcault's September 20, 1896 "Hogan's Alley" strip, which depicts a chaotic scene of dozens of people and pets packed into an urban back street. Westbrook's caption, "The Kid as Anti-Authoritarian," provokes you to consider the cartoon and think for yourself about why he gave it that label. But when you click on the caption, a red square highlights the section of the cartoon that Westbrook finds most revealing—the working-class residents of the alley beating up the dogcatcher, who is seen as a symbol of public authority. A second level of caption leads you to a more detailed analysis of class relations in Outcault's strips.

With such interactivity and hypertextuality, Westbrook goes beyond providing a richer body of evidence than in print, to offering a different mode of argumentation. Considered from a linear perspective, his essay consists of three sections, on "The Business of the Strips," "The Culture of the Marketplace in the Early Comic Strip," and "Spectatorship and Framing in the Early Comic Strip"—with an "appendix" presenting all of the illustrations used in the essay. But conceptually, Westbrook argues that he is doing something else entirely. Each of the three sections, he writes, "approaches the subject matter from a different direction and defends a distinct analysis." Still, the three threads add up to a single essay, "because none of the threads can stand alone," he maintains. "Each depends on concepts and observations built up in the other threads: to make a general argument about the ways that comics work through vital cultural conflicts."

In the case of Louise Krasniewicz and Michael Blitz's "Dreaming Arnold Schwarzenegger," the additional online offerings add up to quite a bit more. The essay deliberately defies easy summary. It is, in part, an exploration of our celebrity culture and Schwarzenegger's iconic place in it and, in part, a discussion of the problem of representation, in dreams and hypertext. As the authors self-reflexively admit, the ultimate "subject matter is us." Such disparate goals explain why the site includes—among other things—descriptions of Krasniewicz's and Blitz's 154 dreams about Arnold; brief comments on at least 18 of his films; detailed essays on two films; 15 magazine covers featuring Arnold; and dozens of 1995 e-mails between Krasniewicz and Blitz discussing love, life, and Arnold. But such an exhaustive (and perhaps exhausting) archive does not exhaust the site, which also contains dozens of links to other Web sites, from Harvard University's Laboratory of Neurophysiology to the Arnold Schwarzenegger Classic Body Building Competition, along with multiple modes of navigation.

If all that seems a bit chaotic, that is, according to Krasniewicz and Blitz, precisely the point. They were attracted to hypertext, they tell us, because the conventional scholarly forms (book, article, conference paper) did not seem to meet the needs of their subject and their analysis. "We needed a medium, a forum," they write, "that would allow us to incorporate not just the more formal components of investigative research, but also the kinds of discoveries and reflections that are more traditionally relegated to the margins of qualitative research." They find in hypertext "a mechanism for connecting disparate information in the same way that a dream does." At least for Krasniewicz and Blitz, hypertext doesn't merely allow them to do a better job of representing the fullness of their work on Schwarzenegger; it is the *only* way of representing it.

Collectively, then, these four essays point up the potential advantages of the new medium for the presentation of scholarship. Still, they are, in the end, more successful as additive than as expressive uses of new media. They give us more. But have they transformed the nature or quality of scholarly argumentation? Castonguay's essay, for example, is rich in multimedia evidence, but it still advances the kind of scholarly argument familiar to readers of film and American-studies journals—an argument about how race, gender, and imperialism were inscribed in film.

Nevertheless, perhaps at some level *more* does become *different*. As Randy Bass points out in one of the three commentaries on the experiment that we published in the print version of *American Quarterly*, the articles alter the traditional scholarly relationship between argument and evidence, between story and archive (particularly in the work of Krasniewicz and Blitz, where it could be said that the archive is the article). For those of us thinking about the future of electronic publication, what may be most important about these essays is that they provoke us to think about the intellectual and practical problems of doing scholarship in cyberspace.

Unless we attend to those practical problems, the "digestion" of new technology is likely to cause considerable indigestion.

One issue is the way that the unsettled state of the technology expands the job of Web authors. They must, for example, become software testers, since pages designed for Netscape on the Wintel machine will look different when viewed with Internet Explorer or on a Macintosh. Authors for the print version of *American Quarterly* are not expected to know page-layout software, while electronic authors must know considerably more than that. They need to become programmers and designers, with technical expertise rarely learned in humanities doctoral programs and the judgment to create an attractive site.

These problems—technical, production, design—do not bother print authors in scholarly journals, chiefly because such journals operate according to a conventional set of standards in layout, typography, and format that were already well established by 1899. In the end, it is the absence of clear standards that makes scholarly work in hypertext both exciting and problematic. Standards are inherently conservative; they are, in part, what make conventional scholarly articles so conventional. But while standards can be deadening, they can also make scholarly articles easy to read, at least by those who know the "codes." Most academics can quickly get the main points of a scholarly article—they can rapidly find the thesis in the first few pages; the conclusions on the last two pages; and a sense of the sources used through a quick scan of the footnotes.

Such reading skills are worthless in confronting the hypertext essays we published. Not only is the thesis hard to find quickly, but it is not always clear that there is a thesis. Where is the beginning? The end? Reader expectations about the investment of time required to master an essay are entirely disrupted. Do you need to read about Krasniewicz's and Blitz's 154 dreams or view all 64 film clips in Castonguay's essay to have "read" their articles? In effect, those works undercut the unwritten social contract that exists between readers and writers of scholarly essays—a social contract in which the author agrees to follow conventions of argumentation, organization, and documentation and the reader agrees to devote a certain amount of time to give the article a fair reading.

And that suggests another issue for electronic publishing. Our innovative hypertexts have been read (or at least visited) substantially less than the more conventional digital supplements that we posted on the Declaration of Independence. Are readers afraid of them?

Evolving new standards and conventions—creating a new scholarly social contract—is not going to be easy. But if we are serious about trying to find ways to do something genuinely expressive with the new media—to create a scholarship that would unsettle J. Franklin Jameson—then we need to attend seriously to developing the next generation of writers (and readers) of scholarly hypertexts.

7

Should Historical Scholarship Be Free?

O N FEBRUARY 3, 2005, THE NATIONAL INSTITUTES of Health (NIH) issued a new policy on "Enhancing Public Access to . . . NIH-Funded Research." It urges NIH-funded researchers to make all their peer-reviewed journal articles available for free to everyone through a central repository called "PubMed Central," within twelve months of publication in a journal. Although the original force of the initiative was diluted through industry lobbying, the NIH measure represents government recognition of the principle that research, especially government-supported research, belongs to the public, which should not have to pay the prohibitively high subscription charges levied by many scholarly journals.[1]

The new policy affects few historians, but its implications ought to give us serious pause. After all, historical research also benefits directly (albeit considerably less generously) through grants from federal agencies like the National Endowment for the Humanities; even more of us are on the payroll of state universities, where research support makes it possible for us to write our books and articles. If we extend the notion of "public funding" to private universities and foundations (who are, of course, major beneficiaries of the federal tax codes), it can be argued that public support underwrites almost all historical scholarship.

Do the fruits of this publicly supported scholarship belong to the public? Should the public have free access to it? These questions pose a particular challenge for the American Historical Association (AHA), which has conflicting roles as a publisher of history scholarship, a professional association for the authors of history scholarship, and an organization with a congressional mandate to support the dissemination of history. The AHA's Research Division is currently considering the question of open—or at least enhanced—access to historical scholarship.

It is the Internet, of course, that has pushed such questions to the forefront because it has broadened access to some historical resources while it has sharply restricted access to others. For the student and amateur, the Web seemingly is a free library and archive, but as most teachers and scholars know, this library has relatively little serious scholarship—especially scholarship in the humanities. If Google's digitization plans succeed, that free archive will get dramatically larger, but the latest historical scholarship will still be absent. The high school student preparing a History Day essay or the history enthusiast researching ancient Rome is blocked at the gates erected by Project Muse, the History Cooperative, Blackwell, ProQuest, and other commercial and noncommercial entities that own or control almost all historical scholarship published in journals. Professional historians routinely complain that their students and neighbors pick up "junk" on the Internet, but they don't adequately consider that the best online scholarship is often only available to paying subscribers. And while their students may have access to this scholarship through the university library, this access is confined to their time in school, as if that proscribed the limits of the public's need to learn from historical scholarship.

Ironically, once scholarly publications have placed their contents online, it actually costs more to maintain the gates that lock out potential readers than it would to open these works to the world. This paradox (along with skyrocketing prices of many scientific periodicals) has fostered the burgeoning movement for "open access" to scholarly work that led to the new NIH policy. In the words of the Budapest Open Access Initiative, one of the founding documents of the open access movement:

An old tradition and a new technology have converged to make possible an unprecedented public good. The old tradition is the willingness of scientists and scholars to publish the fruits of their research in scholarly journals without payment, for the sake of inquiry and knowledge. The new technology is the internet. The public good they make possible is the world-wide electronic distribution of the peer-reviewed journal literature and completely free and unrestricted access to it by all scientists, scholars, teachers, students, and other curious minds.[2]

Should historians embrace "this unprecedented public good"? Should they join growing numbers of scientists in making their scholarship open and free to the public? The advantages of open access are fairly obvious and have been summarized well by key partisans such as Stevan Harnad, Peter Suber, and John Willinsky.[3] They note that journals benefit because their research is, in Suber's words, "more visible, discoverable, retrievable, and useful." Even more important, authors gain greater visibility, a bigger audience, and more impact. A study in computer science finds that online articles are cited more than four times as often as offline articles.[4] Within the History Cooperative, which provides the electronic edition of the *American Historical Review* (*AHR*), the journals that provide open access, such as *History Teacher* and *Law and History Review*, receive more traffic than comparable, gated publications. If, as much evidence suggests, open access increases the readership, reputation, and recognition of authors, scholarly associations devoted to furthering the professional interests of their members need to consider how they can facilitate that enhanced access.

The most important beneficiaries of open access, however, are nonscholarly readers and citizens, who would gain entry to a world that is currently closed to them. Willinsky describes the lack of public access to electronic scholarship as "a secondary digital divide" that "affects health organizations in Indonesia, university students in Kenya, . . . anti-poverty organizations in Vancouver . . . science fair participants in Wichita and high school history teachers in Charleston." "Just as a vast, rich world of information is within a click or two of most phone jacks," he writes, "the toll gates are going up around online scholarly research."

But what does open access mean for scholarly societies like the AHA? That is an important question for all historians, since scholarly societies are much more important publishers of academic journals in the humanities than in the sciences. Whereas two giant companies (Reed Elsevier and Springer) publish 40 percent of the journals in science, technology, and medicine (referred to in the publishing and library worlds by the acronym STM), very few major history journals come from commercial publishers.[5]

Open access to scholarship fits perfectly with the founding principles of scholarly societies. After all, Congress chartered the AHA in 1889 "for the promotion of historical studies."[6] Thus, making the serious scholarly work found in publications like the *AHR* free to every high school student and history enthusiast serves the association's highest goals. But the AHA is a publisher as well as a scholarly society and, as such, giving away the scholarship found in the *AHR* threatens the economic basis of both the association and the journal. If the *AHR* is free, then who is going to bother to pay? And if no one pays subscription fees, then how is it going to keep publishing and how is the AHA going to pay its bills? Indeed, in the sciences, scholarly societies have viewed calls for open access with more suspicion than commercial publishers. The editor of the American Chemical Society's *Chemical and Engineering News*, for example, denounced NIH's open access plan as "socialized science."[7]

Contrary to first impressions, the threat that open access poses to the revenue base of scholarly societies (or other publishers for that matter) has little to do with cancellations by individual subscribers or members. Most members of the AHA and similar societies already have free, online access to the electronic versions of scholarly journals through their libraries. In the long run this free availability may undercut the financial base of scholarly societies and may require fundamental rethinking of their economics, but it is not directly relevant to current open access proposals. The issue is about the libraries themselves. Why would they pay institutional subscription fees—which can be three to ten times the subscription price for an individual—for journals that are free online?

Is there any way to square this circle, to increase access to the scholarship in journals like the *AHR* without putting the sponsoring societies out of business? A great deal of energy in the open access movement has gone into thinking about ways to make that possible. None offers the perfect solution that would both guarantee the financial well-being of scholarly societies and ensure total free and open access to scholarship. But they are worth considering as ways of furthering the larger goal of disseminating scholarly work. Here, very briefly, are six widely discussed approaches:[8]

Self-archiving: Scholars themselves can make their work available for free through personal Web sites, institutional repositories, in disciplinary archives (such as PubMed Central in the life sciences and arXiv.org in physics). Most journals insist on keeping the copyright to scholarly articles, and, hence, authors must seek permission to post the work online (although not "preprints" of their work). Yet, although authors do not always realize it, many journals, including the *AHR*, grant such permission automatically. Even the scientific journal behemoth Reed Elsevier now allows self-archiving by authors who publish in its journals. One obvious problem with self-archiving is that it puts the burden of open access on authors. Another is that it relies on traditional journals, whose finances it might undercut, to do the important work of peer review. Still, the AHA could encourage its authors to self-archive with the institutional repositories (or e-print archives) that are being set up in many research libraries or even create its own history archive. The evidence so far is that long-established e-print archives, such as arXiv.org, with over 200,000 papers, have not affected the circulation of the related journals.[9]

Author charges: Self-archiving means that the original journal itself is still gated. How can journals cover their costs without getting subscription revenue? One much-discussed strategy is charging authors rather than readers. The commercially run BioMed Central (BMC) publishes more than 100 Web-based journals on this basis.[10] Some object that this imposes an unfair burden on less affluent authors—or at least those without institutional or grant support. But BMC waives fees on a case-by-case basis. Others argue for a differential system in which those who pay a set fee would have their scholarship openly available, a policy recently adopted by Springer (although it has set the price at a hefty $3,000).[11] Would such a system succeed in the humanities, where scholars generally lack research grants that support publication charges?

Delayed access: The new NIH policy allows authors to delay the release of their work on PubMed Central for up to twelve months, although it strongly encourages early release. But the delay (which ranges from two to twenty-four months among journals) is offered as a protection to journals on the assumption that libraries would pay for immediate access to important research results. In history, would libraries see delayed access as a sufficient disincentive to canceling subscriptions, especially since our work—unlike much work in the sciences—continues to be read and cited long past the original publication date?

Partial access: The NIH policy applies only to peer-reviewed scholarship; not all of the contents of journals—editorials, letters, and reviews, for example—are included. Applied to a journal like the *AHR*, that would mean only the articles themselves and not the reviews, which take up more than half of a typical issue. If the *AHR* offered free access to its articles but not its reviews, it seems unlikely that libraries would cancel their subscriptions. Of course, gating the reviews means fewer readers and less influence, especially in comparison to the freely available H-Net reviews, which are now linked from the Library of Congress catalog.

Electronic-only journals: One key strategy for opening up scholarship is to reduce the costs of disseminating it. Although some early enthusiasts touted digital publication as "free," almost everyone now recognizes that electronic-only journals have many of the same administrative and editorial costs as print journals. Still, abandoning print does bring significant savings on paper, print, and postage. And some like John Willinsky have been developing innovative open source (freely available) software to automate the management and production of electronic journals, and thereby reduce costs. The degree of savings remains a matter of dispute, but a good guess is that it is around one-third. The difference in startup costs is even greater, however, since new journals don't have any initial revenue and need funds to get themselves established. *World History Connected*, an open access journal on world history teaching, was able to launch much more quickly and much less expensively on the Web than in print. Most of the journals in history with fully open access are, in fact, electronic-only journals. And while such journals still have administrative and editorial costs, sponsoring universities often largely cover those expenses. Of course, many historians shudder at the thought of abandoning print, and certainly the print edition can continue with the delayed open access model, for example, but my guess is that their numbers have begun to diminish significantly as many have begun to pulp their back issues of the *AHR* in the knowledge that they can always get the articles through JSTOR.

Cooperation with libraries: Libraries faced with mounting serials budgets favor open access as a way of reducing these crippling costs. They also object that high-priced serials force universities to "buy back" scholarship that it has already paid for in the form of salaries. John Willinsky proposes that scholarly societies create cooperative ventures in which major research libraries would pledge long-term support for society journals in return for a promise of reduced and controlled subscription fees (achieved perhaps by a switch to electronic-only publishing). "Imagine," he writes, "that 400–500 research libraries worldwide . . . form an alliance to support the publishing programs of scholarly associations at a rate based on perhaps 80 percent of the current subscription fees paid by those libraries to the associations. In return, the scholarly associations would publish their journals on an open access basis. The top research libraries would achieve immediate and long-term savings, while thousands of other institutions would have access to these journals for the first time." Willinsky concedes that such a cooperative "would not be easily achieved," but it is worth exploring.[12]

These proposals range from the incremental to the revolutionary. Some (library cooperatives, for example) would require scholarly societies like the AHA to dramatically alter their ways of doing business; others (such as partial access) would, in my view, have little effect on the association's revenues. Regardless of one's view of the merits of open access (and my own position is obviously in favor of much freer access), these approaches require careful consideration by historians—if only because external pressures (from government, from the rising tide of the open access movement) are likely to force us to reevaluate our policies sooner or later. But the more important reason to consider how we can achieve open access is that the benefits of broad and democratic access to scholarship—benefits that are within our grasp in a digital era—are much too great to simply continue business as usual.

8

Collecting History Online

with Daniel J. Cohen

I TS VERY NAME—THE INTERNET—UNDERSCORES how this advanced computer network exists to shuttle information *between* and *among* people. It does not, like print, merely deliver documents from point A (historians) to point B (audience). If we want to make full use of this two-way street, we must go beyond passive "texts" such as Web sites and Web pages and also think about active processes such as communications and interaction.

To be sure, historians have already largely embraced such activity on the Internet. Almost all of us use e-mail, and an increasing portion use instant messaging and other forms of online communication. Thousands of professional historians participate in the 150 discussion groups sponsored by H-Net.[1] Enthusiasts and amateurs are involved in dozens of discussion boards and forums sponsored by the History Channel and Yahoo. In contrast to paper media, the Internet seems ideally suited for this kind of vibrant, daily exchange.

Another form of interactivity on the Web remains less developed but has the potential to create novel forms of history in the future: using the Internet to collect historical documents, images, and personal narratives, many of which would be lost if historians did not actively seek them out. For historians working on topics in the post–World War II era, the Web can be a valuable yet inexpensive tool for reaching individuals across the globe who might have recollections or materials. Moreover, a significant segment of the record of modern life exists in digital form. Historians will need to find ways to capture such documents, messages, images, audio, and video before they are deleted if our descendants are to understand the way we lived. This chapter explains step by step how to use the new technology of the Internet in service of the ancient practice of collecting and preserving the past.

Why Collect History Online?

Think for a moment of the outpouring of thoughts and emotions in thousands of blogs on September 11, 2001, or the breaking news on the home pages of myriad newspaper Web sites. A large percentage of this initial set of historical sources, unlike paper diaries or print versions, will likely be gone if we look for them in ten years. Blogs disappear regularly as their owners lose interest or move their contents to other systems or sites. Similarly, unlike the pages of their physical editions, newspaper Web sites change very rapidly (almost minute by minute on September 11) and have no real fixity. Had "Dewey Defeats Truman" been splashed across Chicagotribune.com rather than the paper *Chicago Daily Tribune*, it would have taken just a few keystrokes by the newspaper's editors to erase the famous blunder instantly and forever. As we describe here, we felt an obligation to save the rich personal record of blogs in our September 11 Digital Archive so future historians could understand the perspectives of thousands of ordinary people from around the world. Through even swifter action, the Library of Congress, the Internet Archive, WebArchivist.org, and the Pew Internet and American Life Project were able to save thousands of online media portrayals of that day's events. Had they decided months later to save these Web pages, instead of within mere hours, many already would have vanished into the digital ether. Collecting history online may not always be this urgent, but these examples show the critical need for historians to find the most effective ways of using this new technology to supplement the historical record on paper, as we did in the twentieth century with tape recorders and video cameras.

This is particularly true because we can use the Internet for more than just gathering the history that was made online, or "born digital." The Internet also allows us to reach diverse audiences and to ask those audiences to send

FIGURE 8.1   An important, but highly ephemeral, piece of digital history: the home page of the *New York Times* Web site at 4:43 p.m. on September 11, 2001. Stripped down to its bare essentials so as to reduce the stress an exponential growth in news seekers imposed on their server (note the basic, rather than gothic, font), and constantly changing as the day went on—with no paper trail of these many "editions"—the page would be gone forever had the Library of Congress and the Internet Archive not acted immediately to capture it.

us historical materials that originated offline, or at least off the Web. They can "upload" to us their digital or scanned photos, their sound recordings, or their lab notes. They can use a computer keyboard or microphone to transmit to us their recollections of earlier events and experiences, especially ones for which there are no or few records.

Unfortunately, using the Web to gather historical materials is harder than using the Web as a one-way distribution system. It can involve more technical hurdles than a simple history Web site; legal and ethical concerns, such as invasion of privacy and the ownership of contributed materials; and skills, such as marketing techniques that are unfamiliar to most historians. In addition, collecting online elicits concerns about the quality of submissions: given the slippery character of digital materials, how can we ensure that what we get is authentic, or that historical narratives we receive really are from the people they say they are? How can we ensure that a mischievous teenager isn't posing as an important historical subject? Moreover, some historians argue—not without merit—that online collecting excludes those older, less educated, or less well-to-do subjects who may not have access to the necessary technology. They also worry that the nature of such collections will inevitably be shallow, less useful for research, and harder to preserve.

Some of these worries are relatively easy to address. In our experience, for instance, teenagers are generally too busy downloading music to play games with historians and archivists online. But other concerns are not as easily answered. Collections created on the Web through the submissions of scattered (and occasionally anonymous) contributors do have a very different character from traditional archives, for which provenance and selection criteria assume a greater role. Online collections tend to be less organized and more capricious in what they cover.

They also can be far larger, more diverse, and more inclusive than traditional archives. Indeed, perhaps the most profound benefit of online collecting is an unparalleled opportunity to allow more varied perspectives to be included in the historical record than ever before. Networked information technology can allow ordinary people and marginalized constituencies not only a larger presence in an online archive but also a more important role in the dialogue of history.

Furthermore, in contrast to traditional oral history, online collecting is a far more economical way to reach out to historical subjects. For example, because subjects write their own narratives, we avoid one of the most daunting costs of oral history, transcription. Consequently, although live individual interviews are often quite thorough and invaluable resources, online initiatives to collect personal histories can capture a far greater number of them at lower cost, while at the same time acquiring associated digital materials (such as photographs) just as cheaply. Of course, even if highly successful in the future, online collecting will not mean the end of traditional ways of gathering recent history, including what will surely remain the gold standard, oral history. As oral historian Linda Shopes observes, newer technological methods will have a hard time competing with many aspects of the oral historian's craft: "the cultivation of rapport and . . . lengthy, in-depth narratives through intense face to face contact; the use of subtle paralinguistic cues as an aid to moving the conversation along; the talent of responding to a particular comment, in the moment, with the breakthrough question, the probe that gets underneath a narrator's words."[2] Using the Internet will likely sup-

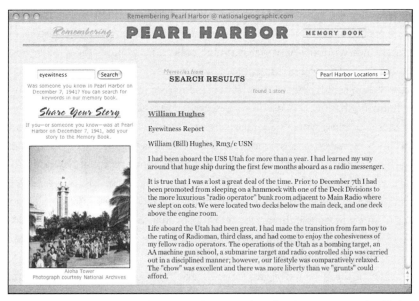

FIGURE 8.2 *National Geographic's* Remembering Pearl Harbor has a "Memory Book" that allows visitors to record first-hand accounts and other recollections about World War II.

plement or complement older, more time-consuming and costly methods such as this.

Despite the pitfalls and insecurities about online collecting, it has become a burgeoning practice. Recently, for example, the British Library, the Victoria and Albert Museum, the Museum of London, and several other British museums and archives have pooled their resources to display and collect stories of immigration to the U.K. in a project called Moving Here. Thus far, the project has posted almost 500 stories and artifacts—mainly digitized versions of existing archive records but also new materials acquired via the site—ranging from a documentary video on Caribbean life to the reflections of recent African immigrants. The British Broadcasting Corporation's two-year online project to gather the stories of Britain's World War II veterans and survivors of the London Blitz, entitled WW2 People's War, has been even more successful, with over 1,000 narratives gathered through the BBC's Web site after only eight months, including dozens of harrowing accounts of D-Day.[3]

In the United States, the National Park Foundation, the National Park Service, and the Ford Motor Company are using the Internet to collect firsthand narratives of life during wartime for a planned Rosie the Riveter/World War II Home Front National Historical Park in Richmond, California. So far, more than six thousand former home front workers have contributed stories. *National Geographic*'s Remembering Pearl Harbor site has received over a thousand entries in their "Memory Book." Over five hundred people have recorded their personal stories and artifacts of the civil rights movement on a site cosponsored by the AARP, the Leadership Conference on Civil Rights, and the Library of Congress. The Alfred P. Sloan Foundation has taken a pioneering role in encouraging more than two dozen online collecting projects (including our own) in the recent history of science and technology, arguing that this history is growing much faster than our ability to gather it through more conventional means. Though there remains a healthy skepticism in the oral history community about the usefulness and reliability of narratives collected online, several new projects by major oral history centers (such as at Texas Tech University) show that they, too, are noticing the benefits of online collecting. Even Columbia University—the home of the nation's first oral history program—is encouraging alumni to join in writing "Columbia's history" by contributing stories online.[4]

Good Candidates for Online Collecting—and Poor Ones

Not every topic lends itself well to an online collecting project. Even with the global reach of the Internet and the world's most interesting subject matter (undoubtedly, whatever you as a historian study), you will need to connect with a fairly sizable body of contributors for your project to succeed. A Web site seeking personal narratives of the Roaring Twenties will fall flat (consider the average age of a person who can recall that era), as will most projects targeting topics before World War II. One Web site, on the history of Greenland ice drilling (co-sponsored by the American Meteorological Society, the American Geophysical Union, and the American Institute of Physics), attempted to capture memories from scientists who have gone to Greenland to study the environment by sampling tubes of ice drawn from millennia-old sheets, but it eventually faced a difficult reality: although the ice drilling projects in Greenland are tremendously important for ongoing debates about critical climate issues such as global warming, the number of climatologists and geologists who have set up and run such experiments is relatively small. There simply are not enough of them to populate the site with a highly active historical discussion.[5] Less obvious than the problems associated with a small pool of potential contributors is the opposite quandary: a topic so broad that it fails to excite any discernable cohort. A project directed at collecting the experience of "senior year in high school," in general, is much less likely to attract participants than one directed at the graduates of a particular high school.

Between too sparse and too diffuse pools of contributors, there are many large but discernable communities that will likely respond well to an online project that solicits, archives, and presents their stories and related images, audio, and video. A good candidate for a collecting Web site often revolves around a topic that already has an active, historically conscious online community. For example, Apple Computer's fanatical user base and committed employees have engendered numerous sites on the history of the Macintosh, including two major efforts to record the firsthand recollections of those who worked at Apple in the late 1970s and 1980s: the Computer History Museum's Apple Computer History Weblog and Apple software engineer Andy Hertzfeld's Folklore.org Web site. David Kirsch's electric vehicle history site appeals to a relatively small but committed, almost cultlike community of enthusiasts who were experimenting with zero-emission cars long before the major automobile manufacturers were. These hobbyists were used to exchanging helpful information with each other over the Internet. Before the much larger effort of the September 11 Digital Archive, we began our experiments in online collecting through similarly focused histories of recent science and technology. Our project, Echo: Exploring and Collecting History Online—Science, Technology, and Industry, was funded, like the September 11 effort, by the Sloan Foundation as part of its program encouraging the use of the Internet to gather history.[6]

Online collection efforts tied to a real-world event, institution, or social network have a good chance of attracting and sustaining involvement. Many school and college alumni associations run discussion boards that recall the glory days. Flourishing sites related to actual communities, like one devoted to Brainerd, Kansas, or the Rowville-Lysterfield community center in Australia, craft online spaces where local people build their own historical record, contributing family histories, reminiscences, folklore, and personal artifacts such as photos and scanned documents. Sites connected to museum installations such as the powerful Atomic Memories site at San Francisco's Exploratorium use the shared experiences of visitors to their physical exhibitions to encourage storytelling and historical reflection online. Established virtual

FIGURE 8.3 To capture the early history of Apple Computer from those who were there, Andy Hertzfeld, one of Apple's pioneering software engineers, set up the Folklore.org Web site.

communities for seniors, including SeniorNet's World War II Living Memorial and the History Channel's Veteran's Forum, host discussion "threads" with thousands of historical recollections and conversations about the past, and they allow veterans to reconnect to some of their most distant yet most significant life experiences and social networks. Indeed, perhaps the most active sites collecting history online today, for example, the World War II Living Memorial and the Veterans' Forum, exist primarily to enable personal connection among their membership. These sites simultaneously engage participants by bringing them into contact online with their cohort, while encouraging them to relate the details of their lives and the times through which they lived.[7]

Topics that do not match up with an existing online community or offline association still have the potential to succeed, but only if they are carefully framed to make them attractive to a discernable body of contributors. Clearly delimited audiences make it easier to target potential contributors and for these potential contributors to feel comfortable in the knowledge that they "belong" at a site. For example, Joshua Greenberg's Video Store Project provided a space for the original owners and employees of video stores, before Blockbuster bankrupted or bought out most of them in the 1990s, to discuss their history. An eclectic bunch, this "invented community"—more than nine hundred people—enjoyed the chance to recall the early years of the revolutionary technology of the VHS and Betamax, and to read the recollections of others. In turn, this online collection helped sharpen Greenberg's sense of themes to highlight in his dissertation on the social uses of video technology and provided him with a cache of primary sources that complemented printed sources well.[8]

Tools for the Online Collector

Once you feel confident that your topic is a good candidate for an online collection effort, you can begin to explore the technologies you will need to do the actual collecting. Use the right level of technology for your project. Not everyone needs a Library of Congress–grade archival system or the capacity to store millions of digital files. Regardless of the size of your project, you should not overlook existing technologies that can make your job easier. Much of the infrastructure and software required to do online collecting has already been built and written, and you should take advantage of these technologies, where possible, rather than reinventing the wheel.

Choosing a contribution mechanism that is comfortable for your audience is critical. For example, if you are gathering young soldiers' experiences of the Iraq War, you may want to consider using instant messaging as a collecting technology. By contrast, World War II veterans might prefer a more "traditional" e-mail correspondence. The National Park Foundation's Rosie the Riveter site has a Web form for contributors to enter their recollections, but it also has a prominent e-mail link to Rosietheriveter@nationalparks.org. Though the BBC mandates that all contributions to their WW2 People's War collection come through the Internet, they have joined forces with more than two thousand public computer clusters (such as community centers and libraries) to help seniors navigate the Web site and type in their entries.[9]

Similarly, you will need to think about what you would like to collect from contributors, and plan accordingly. If you are doing a project on the history of the Chicago Mercantile Exchange, you may want to consider collecting BlackBerry messages, a form of electronic communication very popular among traders; a project on tourism might do better focusing on digital photographs. In addition, you may want to change your collecting technology as the project gathers steam (and contributions), starting out with the simpler technologies we discuss at the beginning of this section and moving later on to more complex mechanisms.

Probably the oldest and still quite useful technology for online collecting is e-mail—the choice of some of the most successful collecting projects. Almost all people who have Internet access have e-mail and feel comfortable with it. Keith Whittle's Atomic Veterans History Project, devoted to the community of veterans who participated in nuclear testing during the Cold War, has collected and posted more than six hundred personal narratives from veterans, acquired solely through e-mail. Furthermore, as Whittle discovered, e-mailers can send attachments such as digital photographs, many of which now grace the site alongside the narratives. E-mail also allows for long-term interactions, follow-up, and detailed exchanges. An online collecting project can get started right away with a simple, static Web design that, like Whittle's, uses e-mail links to encourage and accept submissions.[10]

Another possibly helpful collecting technology, closely related to e-mail, is the listserv. If you work at a university or other large institution, you probably have access to listserv software, which essentially functions as a group e-mail and newsletter distribution mechanism. In addition to personal narratives, primary documents, and exhibits, the Sixties-L discussion listserv, hosted by the University of Virginia, maintains an ongoing active recollection (and scholarly discussion) of that decade of upheaval. Since its inception in 1997, there have been almost five thousand postings.[11]

Web-based collecting mechanisms need not be much more complicated than e-mail. The explosion of online diaries, or blogs, has given millions of Internet users a taste of what it is like not just to read and view the Web but also to add their own perspectives to the medium. Without any knowledge of HTML or databases, historians can use a blog as a dynamic Web site for collecting and presenting the past. Many ways of maintaining a blog allow for more than one person to post there, thus enabling a community of historical participants to create an ever-expanding discussion about whatever topics interest them. Blogs may also allow for the exchange of images, other digital files, and more recently, audio, as well as links to other online materials. Through a modicum of additional Web design, you can integrate a blog with a static site (or simply link to it from your main site and be satisfied with a clashing design) to have both an archive or gallery of historical materials and a way for visitors to post additional materials to the collection. Blogs generally have built-in search features and the ability to roll up what you collect for export to other locations (such as a different server or the desktop of your personal computer).

Contributors can submit to a blog-based collecting project in several ways. They can e-mail their responses to you, and you can then post them to the site. Alternatively, you can share your blog's update mechanisms—the e-mail address for automatic posts and the location of the Web form you use to add entries to the site—and have others post directly, though this would not allow you to vet submissions first. More securely, most of the blog systems also allow you to set up a multiperson blog that permits anyone in a defined group to post materials to the site through individual accounts. This could work particularly well for small numbers of contributors who know each other—for instance, a group of professional colleagues or friends. Using the upload feature of blogging systems, you can also have members of the group send photographs to be archived on the blog. Other file formats are available too; recently Google added the possibility of audio recordings "uploaded" to a Blogger site using a telephone. The ease with which historians can set up blogs and have people add recollections and artifacts makes them an attractive possibility for a simple collecting site.

One disadvantage of blogs, however, is that they encourage stream-of-consciousness writing and are by nature somewhat disorganized. Threaded discussion or forum software is generally better at creating distinct subtopical areas, so that you do not end up with an undifferentiated mass of rambling contributions. This technology is not new; millions of people have posted to the venerable and still quite active Usenet discussion groups that predate the Web. In addition to imposing a higher level of order on contributions, discussion software packages, like more advanced multiperson blogs, allow you to keep better track of contributors because they can be set up to require users to log in and provide identifying information (such as e-mail and other ways of reaching them). There are several easy-to-use forum programs and hosting services. (Like blogging software, these should be installed on your server by a Webmaster or someone with technical knowledge of database software and programming languages.)

The most powerful and flexible way to receive collections is through an interactive Web site of your own design. Most of the blog and forum programs run on databases behind the scenes, and you (or your programmer) can create your own unique collection system from scratch using the same technology. The great advantage of this approach is that it allows you to set up customized Web forms for visitors to enter information and files, as you find on Web e-mail systems like Yahoo Mail or Hotmail. Blog and forum programs normally have a single box for text entries and a rigid set of shorter questions about the contributions or contributor, making it difficult to ask historical subjects a series of questions about their experiences, or to ask more open-ended or evocative questions suitable for drawing out historical recollections. For instance, on its Web form, Moving Here (the site that chronicles immigration to the U.K.) asks contributors to enter the *range* of years in which their historical narrative occurred—the single date box you commonly find on a blog would have been inadequate for this purpose.[12] Customized Web applications allow greater flexibility in presenting contributions as well because you can continually adjust the way entries are pulled out of the database and arranged on the screen, instead of relying on the existing templates provided by a software package.

Although these features make do-it-yourself collection systems attractive, you should first explore simpler, preexisting programs like blogs or forums to see if one of them meets your needs. To offer a little self-advertising, we might suggest you also check out the Center for History and New Media's free database-driven Survey Building application, which quickly and easily builds forms for acquiring historical files, images, and narratives. Because CHNM hosts the surveys, you don't need to have your own server or know anything about programming.[13] Unique systems may work well, but they often require ongoing maintenance, and if the original programmer leaves the project, they can be difficult to update or fix if there is a problem. If you decide to do it yourself, the learning curve for databases and Web programming language can be steep.

New forms of instantaneous communication on the Internet may further expand the toolkit for collecting history online. Millions are now using instant messaging (IM) software, which permits you to communicate in real time with individuals around the globe. Popular IM software such as AOL Instant Messenger, MSN Messenger, Yahoo Messenger, and Apple's iChat permit file transfers as well, so that contributors not only can recollect the past in online text interviews but also can send you related digital materials as they converse. Although they do not have the tonal inflections of a spoken dialogue, these typed conversations do have the considerable advantage of being self-documenting, unlike oral history interviews, which require expensive transcriptions. More recent versions of these IM programs also allow rudimentary (but rapidly improving) audio and video chats as well, which opens up the possibility of a future that is much like the past of traditional oral history. But fast Internet connections are required on both sides of the line for these advanced multimedia features, which restricts the realm of potential contributors to those with high-speed connections. New services that accept voice recordings via a standard telephone line and convert those recordings into a digital format that you can receive via e-mail or through a Web site offer another possibility for audio collections.[14]

Attracting Contributors to Your Site

Your choice of an appropriate collecting technology probably matters less to the overall success of your project than the content and design of your site and effective outreach to the potential contributors. Too many collecting projects have started with high hopes and ended with a frightfully low number of submissions. You should therefore spend more time thinking about how to excite your intended base of contributors than mastering every last detail of Internet communication.

One strategy for attracting contributors is to offer information or materials that will bring them to the site. What ultimately matters in choosing "magnet" content is not so much its exhaustiveness or the refinement of presentation but rather its distinctiveness on the Web. A small collection of compelling or provocative materials, carefully annotated or explicated in some fashion that appeals to the curiosity of your targeted audience, is far more effective than a vast but conceptually murky collection of materials or a set of documents that can be easily found elsewhere on the Internet. For example, the Atomic Veterans site provides its community with the information that it desires (e.g., an up-to-date collection of declassified documents on nuclear tests), and it presents it in a fashion that reflects the veterans' experience (i.e., as it pertains to specific military operations and outfits). On effective collecting sites such as this one, the magnet content is seamlessly integrated with the contributions, leading to a historical collection that is greater than the sum of its parts. Atomic Veterans steward Keith Whittle has also recognized the importance of keeping a site current and fresh to maintain its attractiveness to visitors and contributors. Like Whittle, you should rotate featured items on your home page or highlight the most recent additions to the collection.

Probably the best magnet content on a collecting site is other contributions. This leads to a major paradox, however, and one that will be familiar to anyone who has ever taught a class: just as no one wants to be first to raise his or her hand, no one wants to be first to contribute to an Internet collection. Potential contributors of personal historical materials may be self-conscious, and even the most eager contributor can fall victim to the worry that his or her story or image will attract too much scrutiny as it sits alone on a featured Web page. Visitors with historical recollections or materials to contribute may even visit the site several times, "lurking" as they try to overcome these worries, and many newly launched online collecting projects have enjoyed great peaks in traffic without seeing any corresponding increase in historical contributions. Thus the paradox: to build a collection, you first need a collection; often the only way to attract contributions is with other contributions. A second contribution is always easier to get than a first, and a third is even easier. Once you've collected a few items, it will become easier to collect more, and a kind of momentum will gradually build.

But how do you establish this momentum? You might ask a related online collecting project or physical archive if you can reprint some of their contributions on your Web site until you have established your project. The coalition of museums and archives behind Moving Here plumbed their rich physical collections for suitable materials that could serve as model "contributions," such as the Jewish Museum's transcripts of interviews with Jewish immigrants to London's East End.[15] If you do not have access to existing collections, try to seek out friends, family, and colleagues who are in the target audience for your site to have them "seed" the collection.

Unless you have such helpful contacts, however, you will likely be reaching out to a new community—one that may not know you—and you will have to convince that group of possible contributors that your project is worth their time. This will require marketing and publicity. Historians usually have no training in such matters, but they are especially important for online collection projects. Successful projects devote much, if not most, of their resources to outreach. Potential contributors have to hear about a site, often repeatedly, before they become interested in contributing. Formulating a detailed outreach plan in advance will help you affirm that you can actually accomplish your collecting goals—after all, if you can't think of a way to reach possible contributors, you will likely face disappointment—and provide a quick jumpstart to your endeavor once the Web site is finished.

Probably the first step is contacting potential contributors directly through e-mail, telephone, or postal mail. When Claude Shannon, the father of modern information theory and the mathematics behind key parts of the Internet such as modems, passed away in February 2001, we launched a modest project to collect his colleagues' reminiscences of him. Sending out approximately three hundred targeted e-mails pointing to our Web site, we collected more than thirty detailed accounts of Shannon's life and legacy from a variety of scientists and technologists, revealing new information about Shannon's work at Bell Labs and his enormous impact on such far-flung disciplines as computer science, computational biology, and genetics. We were also able to collect historical materials from people who would otherwise have been impossible to reach without the Internet. As our initial e-mail was forwarded to listservs and online discussion groups, it ultimately reached a group of scientists working in Siberia who, it turns out, had been profoundly influenced by Shannon's work thousands of miles away.[16]

Indirect marketing focuses on reaching possible contributors through their social networks. You should spend some time identifying and contacting the key organizations and institutions most relevant to your historical subjects. Their assistance to you can range from a simple link on a home page to a feature story in their newsletter to a posting to their e-mail list. If your project genuinely interests their members, they will likely help you. You should also spend some time in your contributors' communities, virtually or in the real world. Become a member of a Web forum, newsgroup, or listserv related to your topic. You may want to attend a live meeting, where you can distribute literature about your project (along with its URL or a phone number to reach you). David Kirsch, director of the Electric Vehicle History Online Archive, spent hours posting on online discussion boards and days attending electric vehicle club meetings to become a trusted member of the electric vehicle hobbyist community and acquire the first set of contributions to his Web site.[17]

Although a direct, personalized e-mail to a historical participant or a community newsletter article about your site may be more effective in building a pool of contributors than a colorful ad in a journal, magazine, or newspaper, you should not ignore the potential of a media campaign or a mass marketing approach. If you can tie your project in some way to current events, a well-placed press release can attract media attention and increase contributions. We experienced some success in this regard for a Web site we built with the National Institutes of Health called A Thin Blue Line, where we leveraged the thirtieth anniversary of the creation of the home pregnancy test to collect the popular history of that landmark reproductive technology. The *Washington Post* and other newspapers ran stories about the site because of its timeliness, which led to a spike in site traffic and a smaller bump in contributions.[18]

Combined with high-quality historical materials, a Web site that collects the history of the recent past can become a trusted center for information on the Web and spark media coverage simply by its existence. When the lights went out in the northeastern United States in August 2003, our colleague James Sparrow immediately received calls from the BBC, the *New York Times*, the *Boston Globe*, National Public Radio, and other major media outlets because he had created the definitive site on the history of the 1965 and 1977 New York City blackouts. With scanned primary documents and audio clips, as well as hundreds of stories gathered via the Web, Sparrow's site shows how an Internet project, done well, can successfully collect history once it has achieved a status as the place to go for a particular historical topic. Following coverage in the mass media, the site gathered over a hundred new personal narratives for its archive.[19]

Encouraging Contributions and Building Trust

Most people will come to your site to view contributions or contribute themselves. You should therefore make it as easy as possible for visitors to contribute and to recognize the value of what you have already collected. You will also need to create trust about you, your site, and its mission. The design of your site and the ways in which you convince possible contributors that their submissions are worth saving for the future are at least as important as the technology. Image-heavy splash pages and Flash movies may be very attractive, but first and foremost you should have clear invitations to "Contribute," "Tell your story," "Read the stories of others," or "View donated images." Beyond these signposts, an attractive design, of course, does burnish the reputation of a Web site and makes it more likely to attract contributions.

When contributors find themselves on the correct Web page to add their recollections or upload a digital file, they should face as few hurdles as possible. Even if you feel that the technology is self-explanatory, provide clear step-by-step instructions, and if possible test them ahead of time with potential contributors. Sites that require logins—usernames and passwords that you must register for in advance of your contribution—will almost always receive fewer submissions than those that allow all comers to proceed. This phenomenon is part of a larger tension between sound (and some would argue sane) archival practice and using the Web to collect historical materials and narratives: the more you ask contributors to reveal about themselves, the less likely they will be to contribute. Librarians and archivists relish "metadata"—that is, solid information about the who and what of an accessioned document, such as the name, address, and other contact information of the creator, and exact details about the provenance of what has been donated. Unfortunately, years of spam, online scams, and poor handling of private information by supposedly trustworthy institutions (such as banks) have made most Web users extremely cautious about entering personal data.

This does not mean that you should accept only anonymous contributions. Instead, we recommend making only contributors' full names and an e-mail address or phone number mandatory—some small bit of information necessary to reach a contributor later—and making other information (such as a mailing address) optional. Moreover, you should ask for this personal information *after* they have completed their submission. Many people will volunteer this information; others won't, but you can use the one piece of mandatory information to contact a contributor later to get further metadata for your collection. This flips the normal order of archival acquisition on its head, of course—get the materials first, then learn about the contributor—but it raises your chances of actually getting contributions in the first place. It also may be worthwhile to offer opportunities for contributors to keep their submissions "private," that is, saved in your collection but unavailable for a time to the public, or to offer to remove their names and other identifying information from any public display.

Likewise, it helps to link from your submission page to a reassuring policy statement that details how you will use the information provided. Specify in bold that all personal information will be closely guarded and not shared for any reason without the consent of the contributor. Online collecting projects should formulate these important policy pages early on and attempt to identify all potential legal and ethical problems. By accepting donations from your contributors, you are assuming responsibility for the information they provide you. It may become too easy to think of your contributors simply as subjects of your research, and for this reason you should remember to treat them with respect.

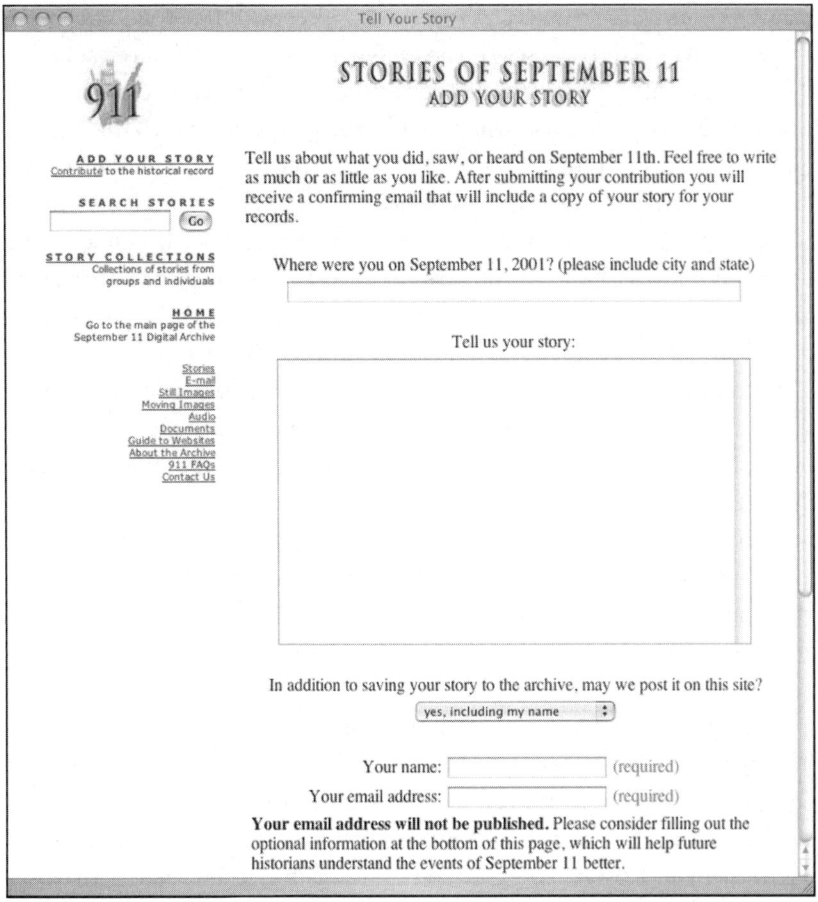

FIGURE 8.4 The stories submission page for the September 11 Digital Archive, which is cosponsored by the Center for History and New Media and the American Social History Project, highlights some of the principles of online collecting forms, including a large upfront box for the first-person narrative, much smaller secondary metadata collection boxes (Zip code, age, gender, and so on), and the importance of building trust, in this case by highlighting our limited use of the contributor's e-mail address.

If you are associated with a college or university, these concerns may have a legal import as well. Because of controversies surrounding medical and psychological experiments, all research involving "human subjects" has come under heightened scrutiny by institutional review boards (IRBs) that oversee university-based research. Whether or not oral history—to which this online collecting can be compared—should fall under the regulation of IRBs remains a controversial subject. Although we agree with those who believe it should be excluded (because it is very different from the kind of research that the federal regulations are directed at), many IRBs disagree, and if that is the case at your university, you will have to have your project reviewed for approval. If so, it will probably help your case if you describe it as "online oral history." Describing your work as a "survey" will place it in a category of social science research where it doesn't belong and where it will fall under closer official scrutiny. Whether or not you need to have your plans officially reviewed, you should always strive to follow the ethical guidelines provided by disciplinary organizations such as the Oral History Association.[20]

Regardless of your affiliation, a well-crafted policy page will help protect you and your contributors from ambiguities. The terms of attribution and ownership must be made completely clear, and participants should indicate their informed consent by acknowledging, through either a button click or a check box, their consent to a set of terms for every submission. These consent forms need not be overly detailed or legalistic. They merely should state in plain language what the rules of submission are, where the contribution is going, what may be done with it (such as transferring it to another institution or to other researchers), and whether there may be any further contact from the project staff following the submission. This last part of your policy statement is important because some people will consider further unsolicited entreaties as annoying spam. Although oriented toward companies, the online watchdog TRUSTe has a helpful guide to crafting such policies, including handling disclosures about personal information and related matters.[21]

The Moving Here site on immigration to the U.K. facilitates contributions and builds trust with contributors extremely well. It has a well-designed, simple entry form for stories, with boxes for the contributor's name and an e-mail address or phone number. A clear note about the need to contact contributors and a reassuring link to their privacy policy, which is admirably less than two hundred words, sits adjacent to the short form. The policies on privacy and data protection are devoid of legal jargon, and they pop up when requested so the contributor doesn't need to leave the entry page. The Rosie the Riveter stories site requires only an e-mail address to proceed, though it gently requests other information such as a contributor's name, phone number, and mailing address. Unfortunately the site puts some people off by mandating that they agree to a long, legalistic "Terms of Submission and Disclaimer" before beginning the submission process, though clearly this has not dissuaded the thousands of Rosie the Riveter contributors.[22]

We suspect that this acceptance is in part due to the reputation of the high-profile institutions behind Rosie the Riveter, prominently displayed with large logos on their Web site. Even if your site is not sponsored by a giant automobile company or government agency, it builds trust with contributors if you scrupulously reveal who you are and where they can find you. You should play up any affiliations because they give a Web site a feeling of being connected to the real world and provide a sense of where the contributions are going. Indeed, you may want to make an alliance with a local library, historical society, or university archive to sponsor your Web site and perhaps even house the final collection (in digital or printed form). Many people still consider the Web an ephemeral medium; knowing that their donations have a nonvirtual home helps to overcome the hesitation engendered by this feeling of impermanency. Partnerships with brick-and-mortar institutions help add "weight" to an otherwise "weightless" online project.

Qualitative Concerns

Once you convince contributors to participate in your site, what do you actually want from them, and how can you ensure that what they submit is useful and authentic? Oral historians, ethnographers, and sociologists have carefully thought out sound and effective "instruments" (controlled and rigorous methods for collecting information).[23] Although consulting this literature is worthwhile, the limitations of Web forms and other collection mechanisms such as e-mail work against the strict replication of these standards online. For example, the common method of repeating an important question several times, using a different phrasing each time, to ensure an accurate answer from a respondent sounds good in the abstract. But the size and resolution of most computer screens permit only a fairly limited amount of text and response space, thus requiring a great deal of scrolling (and frustration) as the number of questions proliferate. Historical surveys on the Internet—in the interest of attracting a large group of contributors—probably should not match the density and complexity of offline versions. And, in any case, social scientists would not consider them true "surveys" in the sense of a scientifically valid form for collecting information.

In your Web form, discussion board prompt, or e-mail exchange, be wary of asking for too much. As Don Dillman, a sociologist who has studied the effectiveness of Web surveys, notes, "Survey designers try to get too much detail from respondents. The result is survey abandonment, which the Internet makes relatively easy." Try to keep at least the initial entreaty short and as open-ended as possible—certainly fewer than ten questions, and probably better under five. We have found that some of the most effective online collections projects involve not much more than a call to "Share your story." (Yes, most people are narcissistic and like to talk primarily about themselves and their experiences.) In the narrative that results from such open-ended questions, you will often find the answers to more specific questions that would have been far down a long and off-putting survey form. One of the online collecting efforts related to September 11, 2001 (see below), simply asked "Where were you?" and yet was able to collect a vast archive of rich firsthand narratives—not just of where, but of when, with whom, and how their diverse set of contributors experienced that day. Historians new to survey design tend to be too specific. Although you may think a detailed survey will get you exactly what you want, it may in fact be confusing to visitors who do not know as much about the broad sweep of your subject as you. In addition, asking questions in the authorial voice of a scholar or book repels most potential contributors. Write questions in a more casual prose style.[24]

Finally, be flexible. Remember that one of the advantages of the Web is its ease of revision: you can change wordings and collection formats at any time if things are not working out the way you envisioned. (Do be sure to save earlier versions so that future researchers can properly understand older sets of responses.) You should also be prepared to accept things you had not intended to collect. Sometimes contributors will want to give you materials you did not ask for or tales that seem unrelated to your focus. Always consider accepting these donations. The public's generosity may surprise you, and it may enrich your project in ways you could not have anticipated.

Of course, along with great generosity sometimes comes undesired mischief. How can you be sure your contributors are who they say they are? How can you be sure their contributions aren't faked, or taken from other sources? Concern about the falsification of digital historical documents and materials, we believe, has mostly turned out to be a phantom problem. We are not alone in this assessment. Newspaper Web sites, which rely on the registration information given by surfers to make money off of targeted advertisements, have found (much to their surprise) that relatively few people enter fake information, even though there are sometimes no checks against such subterfuge. In one study the *Philadelphia Inquirer* discovered that only about 10 or 15 percent of their 300,000 registered users had entered bad e-mail addresses (and some of those were merely by accident or due to technical difficulties), even though a person's e-mail address is among the most guarded possessions of the online world because a vast majority of people are worried about spam. Zip codes and other less problematic bits of personal information are falsified at an even lower rate.[25]

We think the nonprofit mission of online historical archives generally produces even higher rates of honesty. Most people who take the time to submit something to your project will share your goals and your interest in creating an accurate historical record. Rogues and hackers have more interesting things to do on the Internet than corrupt historical archives. But our best defenses against fraud are our traditional historical skills. Historians have always had to assess the reliability of their sources from internal and external clues. Not only have there been famous forgeries on paper; written memoirs and traditional oral histories are also filled with exaggerations and distortions. In the past as in the present, historians have had to look for evidence of internal consistency and weigh it against other sources. In any media, sound research is the basis of sound scholarship.

Nevertheless, some technical methods can help double-check online contributions. Every computer connected to the Web has an Internet Protocol (IP) address. A small bit of programming code can capture this address and attach it to the other metadata associated with a contribution. If you are skeptical that a contribution has come from a specific person or location, a WHOIS search, which translates the numbers of an IP address into a semireadable format that often includes a contributor's Internet service provider and broad area of service, occasionally results in helpful information.[26] Less cloak-and-dagger is a simple e-mail or telephone follow-up with the person to thank him or her for the contribution; if the e-mail bounces back or the phone number is incorrect, you should be more skeptical of the submission. Following up in this way also presents an opportunity to ask contributors if they might have any other documents or recollections and whether they might know of others who can supplement your archive.

A less obvious but perhaps more important measure of the "quality" of a historical collection created online becomes apparent when the collection is assessed as a whole rather than on the level of individual submissions. Like any collection, online or offline, a minority of striking contributions will stand out in a sea of dull or seemingly irrelevant entries. Historians who have browsed box after box in a paper archive trying to find key pieces of evidence for their research will know this principle well, and it should not come as a surprise that these grim percentages follow us into the digital realm. Yet as we also know, even a few well-written perspectives or telling archival images may form the basis of a new interpretation, or help to buttress an existing but partial understanding of a certain historical moment. At the same time, the greater size and diversity of online collections allow you more opportunities to look for common patterns. Why do certain types of stories reoccur? What does that tell you about both popular experience and the ways in which that experience gets transformed into memory?

Moreover, because of a digital collection's superior manipulability compared to a physical collection, historians can search electronic documents in revealing and novel ways. On the Web, the speed with which one can do this sort of analysis can enable both quick assessments of historical collections as well as more substantive investigations. For instance, when historian Michael Kazin used search tools to scan our September 11 Digital Archive for the frequency of such words as "patriotic" and "freedom," he came to some important, if preliminary conclusions about the American reaction to the terrorist attacks. Kazin discovered that fewer Americans than we might imagine saw September 11 in terms of nationalism, radical Islam versus the values of the West, or any other abstract framework. Instead, most saw the events in far more personal and local terms: the loss of a friend, the effect on a town or community, the impact on their family or job.[27] The ultimate quality of a digital collection may have more to do with the forest than the trees, so to speak.

Case Study: September 11, 2001

The most immediate, successful, and helpful examples of online collecting thus far indeed arose in response to those terrorist attacks of September 11, 2001. The date was a watershed moment in the short history of online collecting, a point at which the practice spread in a spontaneous response to historical events among amateur historians, as well as more deliberate responses from professional historians, museums, libraries, and historical societies. We participated in one of these projects, the September 11 Digital Archive, and as part of our project we cataloged hundreds of other sites that were accepting contributions in the form of stories, reflections, artwork, and photographs. To be sure, much of this activity took place in preexisting online locations, such as major media Web sites. For instance, the Web portals of the *New York Times* and the BBC had extensive and quite active message boards recording responses to the events of that day and its aftermath. At the same time, however, many new outlets popped up to collect these feelings and perspectives in a vibrant example of what the Pew Internet and American Life Project has called "do-it-yourself journalism."[28] Starting with the clear conviction that momentous events were occurring around them, scholars, students, archivists, businesses, and members of the general public started online collecting projects in an effort to record the terrible events of September 11 and its aftermath.

Amateur collectors founded many of the earliest successful efforts to capture the history of September 11 online. Wherewereyou.org succeeded in collecting more than two thousand personal narratives of September 11 in the space of just a few weeks.[29] Working quickly and with admirable technical and design skill, the creators of the site developed a database-driven application that allowed people from around the globe to tell their story of September 11. Remarkably, the entire project was unfunded, and conceived and executed entirely by three undergraduate college students working in different cities in their spare time. Wherewereyou.org shows how the Internet can empower amateur historians who want to collect history.

In addition to scores of such amateur efforts, several large-scale professional and institutional efforts used the Web to capture historical materials and narratives. Building on a partnership with the Internet Archive forged to collect Web content during the preelection months and in the aftermath of the contested 2000 presidential contest, the Library of Congress's Library Services Directorate moved immediately to capture Web content related to the attacks. Their September 11 Web Archive officially launched on October 11, 2001, though the Internet Archive's computers began scanning the Web just hours after the September 11 attacks. Led by the library and funded with a grant from the Pew Charitable Trusts, this effort, like the project surrounding the 2000 presidential election before it, sought to "evaluate, select, collect, catalog, provide access to, and preserve digital materials for future generations of researchers." By the time the collecting wrapped up on December 1, 2001, the library and the Internet Archive had collected the contents of nearly 30,000 Web sites and 5 terabytes (5,000 gigabytes) of information, representing an unprecedented snapshot of the world's real-time response to the tragic events.[30]

A second major online effort to document the history of 9/11 was our September 11 Digital Archive, a joint venture of the American Social History

Project/Center for Media and Learning at the City University of New York's Graduate Center and the Center for History and New Media (CHNM).[31] Funded with a grant from the Sloan Foundation, the archive set out to collect, preserve, and present a range of primary sources, especially those born-digital materials that were not being collected by other projects like the September 11 Web Archive. Whereas the Web Archive aimed at collecting public Web pages, our effort sought to collect—directly from their owners—those digital materials not available on the public Web: artifacts like e-mail, digital photographs, word processing documents, and personal narratives. We also wanted to create a central place of deposit for the many and more fragile amateur efforts already under way. Now, more than three years into the project, the September 11 Digital Archive has collected more than 150,000 digital objects relating to the terrorist attacks, including more than 35,000 personal narratives and 20,000 digital images. In September 2003, the Library of Congress formally agreed to ensure the Digital Archive's long-term preservation.

Despite its large scale, the Digital Archive began fairly modestly. To get the site up extremely quickly, we ported over the basic database infrastructure and programming code from several earlier collecting projects on the history of science and technology (our Echo project). Funding from the Sloan Foundation arrived on January 1, 2002, and we launched the site on January 11, with the initial ability to collect digital images, e-mail, and stories. As time went on, we added features as needed, including uploads for digital files other than images and fully automated voicemail contributions. To seed the contributions area, we first publicized the site to friends, family, colleagues, and the students and staff of our respective campuses. On the six-month anniversary of September 11, on March 11, 2002, we had a full public launch with press releases and some major media coverage.

With the technical concerns in the background, we focused heavily on outreach to both a wide audience and the communities near the crash sites in lower Manhattan; Arlington, Virginia; and Shanksville, Pennsylvania. Our marketing efforts paid off over many months as the number of contributions snowballed, and as we were able to forge alliances with the Smithsonian Institution's National Museum of American History as well as other museums and historical societies. Looking at just one section of the site, the one that accepted personal narratives, we had 28 submissions by the end of January 2002, 328 by the end of March, 693 by May, 948 by July, and 1,624 by August 2002. As media attention increased in the period just before the first anniversary of September 11, with major stories on the project on CNN, MSNBC, the Associated Press, as well as hundreds of newspapers, our numbers went exponentially higher. On September 11, 2002, alone we received more than 13,000 personal stories, including hundreds from direct witnesses of the events.

That last note raises a critical second point: these efforts and the growth of the archive led to both a broad response from around the country and the world, as well as from particular audiences we were especially trying to reach. Surely it is easier to garner a general public response via a Web site than to reach a small set of targeted contributors. But we found as the project grew that the sheer number of contributions from the general public made it much easier to gather materials from those directly involved with the events of September 11. Because of our prominence and partnerships with major institutions, we found that key constituencies had heard about us even before we contacted them and were very interested in contributing important historical materials to us, or, if they had not heard of our project, they were much more willing once they went to our site. We had achieved a sort of "presence" and "critical mass" that led to a greater and greater number of contributions and some valuable acquisitions, such as the real-time electronic communications of a group of co-workers evacuating lower Manhattan. Other groups, including Here Is New York, which gathered thousands of stunning photos of the city in the aftermath of September 11, asked us to serve as the repository of their own collections.[32]

The explosion of online collecting following September 11 was part of a larger change in Internet culture that the attacks precipitated. As the Pew Internet and American Life Project has shown, more and more people turned to the Internet as a "commons" after September 11; it became a place to communicate and comment rather than just surf for news. Although most Americans still got their news through traditional media outlets such as newspapers and television and overall Internet usage actually declined in the days immediately following the attacks, an unprecedented number of people used the Internet to share their feelings and perspectives on the tragedies. For example, nearly 20 million Americans used e-mail to rekindle old friendships after September 11. Even more pertinent to the present discussion, 13 percent of Internet users participated in online discussions after the attacks. This interactivity represented an entirely new role for the Internet as a place for community-making and spontaneous documentation. "For the first time," wrote one electronic newsletter editor, "the nation and the world could talk with itself, doing what humans do when the innocent suffer: cry, inform, and most important, tell the story together." More specifically, people approached the Internet as a place to debate the United States government's response to terrorism (46 percent), to find or give consolation (22 percent), and to explore ways of dealing locally with the attacks and their aftermath (19 percent). Such usage of the Internet will only grow in the years to come.[33]

Surely not every collecting Web site will have the scale or results that the September 11 projects have had. Nor should they; not every historical project has a universe of possible contributors equivalent to that of the September 11 Digital Archive. Regardless of size, however, the payoff can be tremendous in a successful online collecting project. The massive capacity of the Web means that historians can push beyond the selectivity of paper collections to create more comprehensive archives with multiple viewpoints and multiple formats (including audio and video as well as text). Given the open access of the Web, it seems appropriate to cast the widest possible net (as it were) in projects like the September 11 Digital Archive, rather than focus on figures such as government leaders who will almost certainly dominate coverage in print. These archives, we hope, will partly make up for their lack of a curator's touch by their size, scope, and immediacy. The nature and extent of what you can gather, though clearly different from a traditional oral history project or museum effort, may be just as enlightening and important as a future historical resource and likely will grow more so as an increasing percentage of our communications and expressions occur in digital media.

Upon reflection, it appears that these online collections of the future are not unlike the very first history of Herodotus, with the potential to promote an inclusive and wide-ranging view of the historical record. In his travels around the Mediterranean region, Herodotus recorded the sentiments of both Persians and Greeks, common people in addition to leading figures, competing accounts, legends as well as facts. He wanted to save all of these stories before they were forgotten so that the color of the past would not be lost. And as he told his audience, he was also cataloging and recounting it all because in the future people might have different notions of what or who is important: "I will go forward in my account, covering alike the small and great cities of mankind. For of those that were great in earlier times most have now become small, and those that were great in my time were small in the time before. Since, then, I know that man's good fortune never abides in the same place, I will make mention of both alike."[34] Using the Internet to collect history shares this vision: it is undoubtedly a more democratic form of history than found in selective physical archives or nicely smoothed historical narratives, and it shares democracy's messiness, contradictions, and disorganization—as well as its inclusiveness, myriad viewpoints, and vibrant popular spirit.

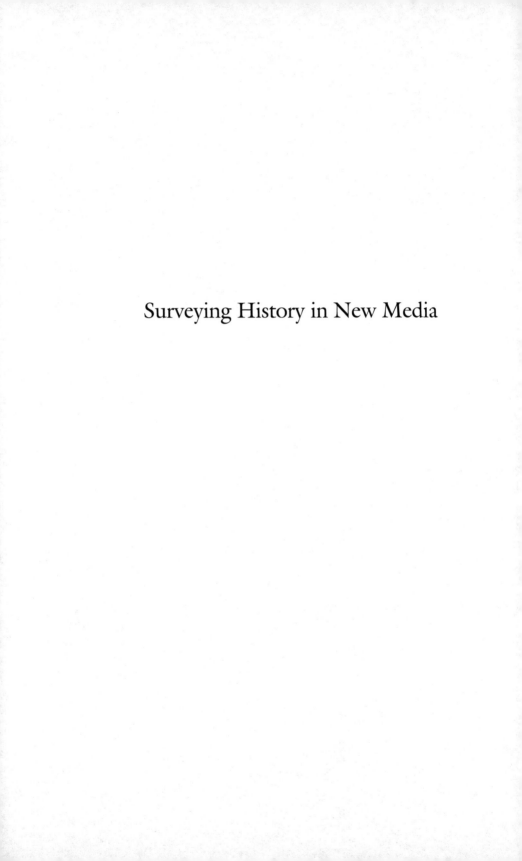

Surveying History in New Media

9

Brave New World or Blind Alley?

American History on the World Wide Web

WITH MICHAEL O'MALLEY

I N AUGUST 1995 NETSCAPE COMMUNICATIONS
Corporation went public at $28 a share; that fall, it briefly reached
a peak of $174—an incredible figure for a company making no real
profits and whose best-known product was essentially free. Even at year's
end, when the share price settled around $130, its market capitalization was
more than $5 billion—greater than the combined market value of the New
York Times Corporation and United Airlines. Netscape's skyrocketing stock
price reflected the sudden discovery by investors and the general public of the
Internet, the global network of connected computers that communicate with
each other by following a common set of protocols. In November 1969 the
Internet's predecessor, the Arpanet (named after its funder, the United States
Department of Defense's Advanced Research Projects Agency), consisted of
just two specially designed communications computers located in Los Angeles and Palo Alto, California. Its initial users were scientists and technical
people, particularly those with Defense Department connections. But in the
1980s and 1990s the Internet rapidly became a broadly accessed medium that
began to rival the telephone and post office in importance.[1]

Even more responsible for the investor gold rush into Netscape stock was
the still more recent emergence of the World Wide Web, whose origins go
back to the efforts in the late 1980s of Tim Berners-Lee, a computer scientist

at the European Particle Physics Laboratory in Geneva, to find a way for physicists to share information easily. The Web, as it is commonly called, uses the Internet's global network, but with a more specialized set of protocols. These protocols make it possible for computers connected to the Web to display pictures, sound, and film; for users to move very rapidly from one Web site to another; and for each Web page to have a single, distinct address (called a "Universal Resource Locator" or "URL"). The result makes the Web into a global "hypertext"—a dynamically linked set of documents or texts.

With hypertext (and the Web), it is as if while reading a book of history, you could click on a footnote and immediately find yourself reading the book mentioned in the note. If the Internet constitutes all the roads of the global computer world, the World Wide Web encompasses its paved roads. To "read" the resources on the Web, you need a Web "browser." Indeed, the release in 1993 of Mosaic, an easy-to-use graphical browser developed by the National Center for Supercomputing Applications at the University of Illinois, first brought the Web to public notice. A year later, in the pattern of rapid commercialization that has characterized the computer industry (with initial funding most often provided by the United States Department of Defense) some of the designers of Mosaic created the Netscape Communications Corporation, whose browser, Netscape Navigator, seemed to be emerging as the de facto standard by the fall of 1995—a fact that made the founders instant millionaires.

There were other indications of Web mania in 1995. In June 1993 there were only 130 Web sites in the world; by June 1995 there were almost 23,500; by June 1996 more than 200,000 new Web sites had come online. Soon, entities from multinational corporations to junior high school students were posting their own "home pages," which can be best understood as the tables of contents or title pages introducing a set of other Web pages. In a prominent front-page article in November 1995, the *New York Times* (perhaps worried about its own eclipse) announced the Web's arrival as a major "social, cultural and economic force" comparable to the "print and electronic media that have preceded it."[2]

Nothing could live up to the Web's advance billing. Just a year later even Wall Street investors had lost their Web fever. The stock prices of some Internet firms, skyrocketing six months earlier, dropped dramatically. One of them, Excite, went public in April 1996 at $17 a share; on its first day of trading it went to $21.25, and by late October it was down to $7 a share.[3]

Skepticism about the Web was not confined to Wall Street. In the same week when Wall Street was decidedly unexcited about Excite, the historian

Gertrude Himmelfarb offered what she called a "neo-luddite" dissent. She was, Himmelfarb wrote in the *Chronicle of Higher Education*, "disturbed by some aspects of . . . the new technology's impact on learning and scholarship." "Like postmodernism," she complained, "the Internet does not distinguish between the true and the false, the important and the trivial, the enduring and the ephemeral." Internet search engines "will produce a comic strip or advertising slogan as readily as a quotation from the Bible or Shakespeare. Every source appearing on the screen has the same weight and credibility as every other; no authority is 'privileged' over any other."[4]

Conservative critics such as Himmelfarb are, in some ways, the inverse of the Web's greatest promoters; what the former fear, the latter welcome. These "techno-enthusiasts" offer, literary critic Randy Bass notes, a new version of the "technological sublime," in which

worldwide connectivity will eradicate physical and political boundaries; . . . the leveling nature of online interaction as well as the universalization of information access will foster democratization; . . . the decentered nature of hypertext will further erode the existence of limiting hierarchies; and . . . the engaging power and linking capabilities of multimedia will revolutionize learning and eradicate the need for teachers and schools altogether.

Hypertext novelist Michael Joyce rhapsodizes about: "the voracious newness of the webbed electronic age with . . . its succession of brave new worlds generated at twenty eight thousand baud and recreated at will." Both cyberenthusiasts such as Joyce and cyberelegiasts such as Himmelfarb put us at the dawn of a new era; the only question is whether it is a utopia or a dystopia.[5]

Other critics argue that the Web marks no departure at all. They dismiss the Web as "just a bunch of links" or "mostly junk." More thoughtful critics on the political Left such as Herbert I. Schiller worry that the Internet will foster and reinforce "information inequality." They argue that inadequate equipment or increasing access fees will shut out the poor. Schiller also warns that the information superhighway may turn out to be "the latest blind alley" if the "tide of commercialism" and its "corporate custodians" engulf the new media and technology as they did radio and television.[6]

This essay offers a preliminary assessment of the possibilities and limitations, the allures and dangers, of the World Wide Web for those interested in presenting, teaching, and learning about American history. The authors are dubious about claims that the Web is a totally new departure. But we also reject the view of skeptics who say that it offers nothing at all; we are impressed—even astonished—by what already exists there for historians. It seems less likely that the Web presents a radically new paradigm or way of thinking; in many ways the Web simply gives us speedy access to existing resources. Yet the very ordinariness of the Web turns out to be interesting; on the Web the past is deeply embedded in the present in ways that escape our notice in the conventional archive or library. Moreover, the power to access information at great distances and great speeds offers the possibility of making new connections—between disparate ideas and between the past and the present—that might otherwise be missed. Finally, the Web offers one key departure—it lets users produce their own versions of history and place them in a public context where no one regulates access, no gatekeeping organizations police content or methodology. We hope to make both the advantages and the disadvantages of this "democratization" more apparent.

This tour of the "History Web" must be brief and highly selective. About sixteen months ago, when we put together a "beginner's guide," it seemed like a "walking city," where users could wander leisurely through and meet all the residents and merchants.[7] Today, reviewing the History Web is more like writing a guide to twentieth-century New York; who can know every street and back alley, who might not miss some of the greatest treasures or the worse eyesores? In this essay, we introduce some exemplary sites and the names of some good guides, hoping that readers will head off and do their own exploring.

We begin with a discussion of how to search for historical information on the Web. Then we offer our own mapping of the Web, organized by types of sites—archives and libraries that have been placed online; "invented archives" (sites devoted to collecting and making available documents that are scattered in various "real" archives), and narrative presentations of history organized by museums, commercial ventures, and amateur enthusiasts.

What's on the Web? Searching for the Past in Cyberspace

How do we find the past in "cyberspace"—the "virtual world" where computer communication occurs? We can start with the Web's many "search engines." These search engines use "Web crawlers"—computer programs designed to follow Web links. They move from link to link, from Web page to Web page, with a mindless tenacity, reporting back all or some portion of all the text they encounter. At the "interface" end, Web page designers and programmers come up with simple ways for users to "query" that filed information. The user's query sends out a search across all the text the Web crawler has found, and the results come back to the user in the fairly sphinxlike shape of "hits." None of the search engines offers anything like the precision, hierarchy, or contextualization historians have come to expect from paper or online library catalogs. The openness of the Web means that the job of cataloging and searching is considerably more complex than that faced by librarians. Search engines are the best available tool, but they cannot tell one page from another or rank the number of hits in any but the crudest way. They cannot tell political Left from political Right, freshmen from grad students, professors from plumbers. But with a little creativity and persistence, one can use search engines to turn up a great deal of information, not all of it what we might expect.[8]

Search engines offer two major approaches—topical and keyword searching. Yahoo! remains the most comprehensive topical directory. If you follow its flow chart from "Arts and Humanities" to "Humanities" to "History" to "U.S. History," you find 873 sites related to American history, broken down into such subtopics as "17th Century," "Museums and Memorials," and "American Flag."[9]

This topical organization lacks any qualitative ratings or guidance beyond occasional brief annotations. On the main Yahoo! U.S. History page, the link to Michael A. Hoffman II's Campaign for Radical Truth in History (a site devoted to racist and anti-Semitic fantasies about the slave trade and the Holocaust) is just one item away from the National Women's History Project. The "Great Depression" offers you a site on the Federal Theatre Project consisting of a single student paper as well as one that presents the Library of Congress's collection of 2,900 life histories from the Federal Writers project of the Works Progress Administration (WPA).[10] Both share equal prominence at Yahoo! This lack of hierarchy between the traditionally sanctioned and well-funded Library of Congress and the unfunded and unsanctioned undergraduate or lone lunatic is one of the most exciting and most unsettling features of the Web.

Efforts at qualitative filtering have emerged. Some of these—particularly the ubiquitous awards to pages as "the best 5% of the Web" or "the best of the Web awards"—need some filtering and qualitative evaluation themselves, since they offer little basis for their judgments. One rating service, Cyber Teddy's Top 500 Web Sites, includes the History Channel and a site on lynch-

FIGURE 9.1 The authors' Center for History and New Media World Wide Web page includes a selection of sites useful to historians.

ing (African American Holocaust) but omits many conventionally important history sites. CyberTeddy explains neither his identity nor his motives. Sites organized by such academic groups as the American Studies Crossroads Project and our own Center for History and New Media provide limited annotation and guides to recommended sites, but none of them has the large staff of a search engine company, which could develop more comprehensive guides or ratings. Indeed, some Web enthusiasts see evaluation and filtering as antithetical to the Web's democratic and anarchic tendencies. Nevertheless, as familiarity with the World Wide Web increases among professional historians, more efforts at filtering will probably develop, but new sources of funding will need to be found. Commercial award pages and the search engines have their own revenue source—advertising. The quest for advertising explains why there are currently several competing search engines; each aspires to become the Yellow Pages of the Web with the lucrative advertising that would presumably go with that status. The Web site for the McKinley Magellan Directory, for example, describes a "picky editorial staff" that awards ratings and then adds the URLs of rated sites to the directory, but it says nothing about what the staff picks and why. Magellan includes useful, brief reviews and ratings of 400 history sites; yet its 65 "four star" sites in history include a page devoted to the musical *Miss Saigon* but not the much more impressive sites from the Library of Congress and the University of Virginia discussed below.[11] The staffs of such commercially oriented projects have little incentive to rate sites with any historical rigor, and so they remain a blunt instrument.

Keyword searches are even blunter, but also more powerful, instruments, as some sample searches reveal. Some topics, events, and people appear more often than others. A search for the Battle of Gettysburg using the AltaVista search engine comes up with about 600 hits; a search for the Homestead strike yields only about 60. Some antielitist figures fare well on the Web. Alta Vista gives us about 700 hits on the radical Emma Goldman but only 64 for the conservative William Graham Sumner and 200 for the Civil War general George B. McClellan.[12]

The search engines combine stunning comprehensiveness with stunning inefficiency. The AltaVista search for Goldman will lead you to the valuable and authoritative Emma Goldman Papers Web site, but it appears as item 122, after you have checked out a newsgroup posting about programming in the computer language C (from a person who includes a quotation from Goldman in his signature line) and a pointer to the home page for the film *Sunset Park* (because one of the stars played the part of Goldman in a workshop production of *Ragtime*). You can find the Emma Goldman Papers site right away if you know that is what you are looking for, but a student or a person simply interested in Goldman might not know it exists. Even if you search for "Emma Goldman Papers" as a phrase combined with Berkeley (where the project is housed), you get 55 hits. "Emma Goldman Papers Project" brings it down to 18 and puts what you want at the top of the list.[13]

Such eclecticism is typical. The 300 pages that turn up in an AltaVista search on Eugene Debs include an *Encyclopedia Americana* entry; rare book catalogs (with books by and about Debs cited); high school and college course syllabi; the home page of a political button collector; a Web site on "noteworthy Hoosiers"; historian Roger Fagge's article "Eugene V. Debs in West Virginia, 1913" in *West Virginia History*; a detailed guide to the Debs papers at Indiana State University; and a request from someone looking for information on the Eugene Debs Sunday Schools that her grandmother attended.[14]

These results are enormously valuable and also incredibly limited. Web readers of Debs's life would come away with some of the basic narrative as well as leads to important resources. They would not read Nick Salvatore's prizewinning biography of Debs; indeed, only if they looked very carefully would they discover its existence. But the students who spent twelve hours exploring Debs's life on the Web would learn something that they could not get (or get as well) from reading Salvatore's fine biography—namely how Debs fits into contemporary American life.[15]

The Web offers an instant education on the uses of the past in the present. Students who explored Debs on the Web would learn that many groups (including some with different agendas and views) claim his legacy—the Democratic Socialists of America, the neoconservative Social Democrats, USA; the Industrial Workers of the World; the Socialist Party USA; and the National Child Rights Alliance—and that they offer different narratives of Debs's life and legacy. They would also learn that Debs is important to individuals from Dominic Chan, who declares himself "an activist and a troublemaker in the proud tradition of Eugene Debs, Mother Jones, Joe Hill and Martin Luther King," and Bernie, who describes himself as "a semi-retired drug dealer," to Noam Chomsky, Ralph Nader, and Cornel West. For anyone interested in how the past is used in the present, the Web is a unique resource. It can allow fascinating assignments that illustrate to students that the past is not dead and forgotten but actively and diversely used.

Searching for Debs's contemporary Andrew Carnegie also offers lessons about the uses of the past in the present. The more than 2,000 sites that a comprehensive HotBot keyword search produced ranged from a site on copyright giving only the dates of his birth and death to an elaborate tribute page with sound clips of Carnegie reading from his essay *The Gospel of Wealth*. This page, part of an online exhibit sponsored by the Carnegie Library of Pittsburgh, showcases some of the Web's strengths. It includes primary source photographs, newspaper articles, editorials by Carnegie, and links to other information. Students would find the primary sources very useful in assessing

Carnegie's philosophy, but the Web site itself is unvaryingly hagiographic, and it does not link the user to any sites that are not equally delighted with Carnegie and his legacy. The rest of the hits tend toward the same. For example, HotBot offers users a connection to the Carnegie Club, a golf club operating in Skibo Castle, once Carnegie's home. The people at Skibo, not surprisingly, have only good things to say about Andrew Carnegie.[16]

The majority of the sites come from the many institutions Carnegie's money made possible—libraries, the famous concert hall, the Carnegie Corporation of New York, the Shadyside Bed and Breakfast. They too do not bite the hand that fed them. On at least a dozen sites, individuals have posted one or more of Carnegie's inspirational essays on success, mostly in connection with free-market ideology. Students will find little balance, little historical depth, and little more than laudatory bromides, and they will learn that Carnegie's money seems to have enabled him to control his representation in history, at least on the Web.

Such fatuous reproduction of received cliché is exactly what some theorists hoped the Web would avoid. Yet, so far at least, the Web—for better *and* worse—seems most interesting precisely for the way it reproduces in digital media crucial features of the "real world." What is *better* is the easy access and fast searching that digital media allow, which instantly highlight connections between the past and the present that historians do not always fully develop in their work. In the same fashion, when libraries and archives come online, the same access and speed can highlight intellectual connections students and scholars might otherwise miss.

Libraries and Archives Online

In a few years, the most important bodies of knowledge for some fields of study will probably be available online and in hypertext through the World Wide Web. But for historians and other scholars who care deeply about information generated before, say, 1990, the vision of a totalizing Web of knowledge will probably never materialize. The chances that the *Worcester City Directory* or issues of *Sound Currency* monthly—sources crucial to our own historical research—will become available online seem remote. Even with vast technological improvements, the costs of digitizing and storing electronically the 110 million items in the Library of Congress are staggering.[17]

However, tens of thousands of those items have already come online through the library's National Digital Library Program (NDLP), also known as American Memory, which began in 1989 and now includes seventeen major Web-based collections.[18] The depth, range, and diversity of these online collections dwarf anything else available for American historians on the Web. The NDLP collections include multiple media (books, manuscripts, films, and sound recordings), but especially photographs; they contain about 70,000 images in eight different collections, from nineteenth-century daguerreotypes to color photos taken by the Farm Security Administration and Office of War Information in the 1930s and 1940s. They also include multiple perspectives—dissidents are well represented in the Woman Suffrage and African American pamphlet collections, whereas such establishment figures as John D. Rockefeller literally have their say in the Nation's Forum sound collections. The Founding Fathers make their appearance in 174 broadsides from the Continental Congress and the Constitutional Convention, while hundreds of unfamous Americans tell their stories in the WPA life histories. The Thomas Biggs Harned Collection of Walt Whitman notebooks offers insights into a now-canonical writer; the American Variety Stage collection, which includes 390 English-language and 80 Yiddish-language playscripts, provides a window into the nation's popular culture.

Despite these riches, the seventeen collections represent a tiny fraction of the Library of Congress's vast holdings. A search for Mother Jones, for instance, turns up just one passing mention in the WPA life histories. But if you are prepared to fit your research agenda to the available collections, you are in much better shape. Students—even graduate students—could write first-rate research papers on such topics as popular photography, conservation, and vaudeville based solely on these collections.

When it decides what to digitize, the Library of Congress considers the material's usefulness to the nation's schools, uniqueness, and appropriateness for Internet formats (maps, for example, are much harder to put online than photos). Copyright status heads this list; documents still under copyright cost too much to publish. American Memory's coverage of the last seventy-five years sticks to either government documents (such as the WPA life histories and the Farm Security Administration photos) or private collections donated to the library with few restrictions, such as the Carl Van Vechten photographs. In the end, financial concerns will most sharply limit what goes online. Reaching even the current goal of 5 million items by the year 2000 will be enormously expensive. In this era of privatization, the NDLP promises to raise $3 in private funds for each $1 in federal money. Reuters America recently donated $1 million for digitization of the papers of George Washington and Thomas Jefferson. Will private funders be as willing to digitize the Emma Goldman Papers as the George Washington Papers?[19]

Still, the advantages of this virtual library stand out. High school students in Florida, college students in Oklahoma, and graduate students in Buenos Aires now have access to an incredible research resource. To be sure, a private high school in a wealthy suburb such as Grosse Pointe, Michigan, is much more likely than a public high school in south Chicago to have functioning Internet access and computers capable of displaying Web pages. Similarly, Internet connections are more plentiful in Western Europe than in Africa and Latin America. Still, the costs of providing adequate access to the Internet and the Web continue to drop. In the United States, a $1,200 computer, a phone line, and a $20-per-month Internet access charge can bring the Web to a high school classroom. We need to remain vigilant lest the Web reinforce the gap between information haves and have-nots. But a Web connection costs less than a good library, and it offers much more than a bad one, an important advantage in an era of declining public funds.

Another advantage of digitization is full-text searching, which allows comprehensive research in online collections. Take the topic of Central Park, a subject on which one of us had spent years doing research at the Library of Congress. In twenty-three seconds of searching in the Library of Congress's American Memory collections, we found 253 references, including some we might never have encountered except by pure chance.[20]

Full-text searching also allows novel methods of intellectual and cultural history research. In this way, the Web's "quantitative" advantages of speed and access can have a "qualitative" (or intellectual) effect on research and teaching by permitting scholars and students to make new intellectual connections between the past and the present and among disparate bodies of material. Suppose you were interested in how Americans in the 1930s used concepts of class, race, ethnicity, or nationalism. The WPA life history collection makes it possible to search for such words as "class," "nation," "American," "black," "Negro," and "Italian," and then to read the interviews in which they appear and see the contexts in which they are used. One of the authors asked his students to pick a subject, enter it into the life history search page, and see how differently people treated that subject in the 1930s. This simple exercise exposed them to some of historians' most familiar problems—that people used different subject categories in the past, or that things that matter

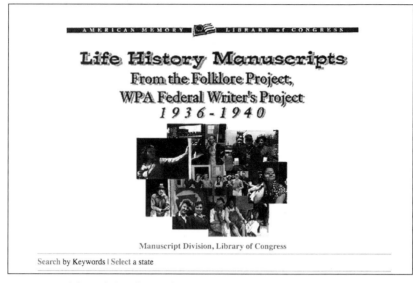

FIGURE 9.2 Through the Library of Congress's American Memory project, users can search 2,900 Works Progress Administration (WPA) life histories for any word or subject.

a great deal to us or that seem like common sense appear very differently. One student found fourteen documents containing the words "gas station" and produced a paper concluding that unlike gas stations today, a small-town gas station in the 1930s functioned as a combination of informal town hall and poor man's social club.

With few exceptions, the quality of presentation for the American Memory materials is high. The project has tried to represent faithfully the original documents, sometimes at the expense of legibility.[21] The visual quality of some photographs (for example, the 25,000 Detroit Publishing Company photos) is disappointing, but the Civil War photos and the daguerreotypes come at much higher resolutions, and in all cases, the viewer first sees a smaller thumbnail version of the photograph, with the option of waiting (under a minute to download over a 28.8 modem, the fastest now commonly used by home computers) for a more detailed view. Some of the American Variety Stage section of American Memory inexplicably comes to the user in an extremely awkward form. For example, users interested in the English Language playscripts will be enticed by titles like *A Mail, a Mick, and a Ford*, but like the other playscripts and most of the Houdini collection, this 1914 "farce comedy" reaches them in the form of "page images." This format requires finding, downloading, and paying for a separate program called an image browser. This cumbersome procedure violates both the norms of the Internet and the library's public mission, and it limits the usefulness of these documents.

The speed of copper phone lines proves a more serious barrier for American Memory's film collections. Over a 28.8 modem, you will have to wait from thirty-five to fifty minutes to see a one-minute clip of President William McKinley's funeral train arriving in Canton, Ohio. Even with a much faster connection, you may wait from five to fifteen minutes for the train to arrive at your desktop. Technological limitations and glitches still hinder the user, but on the whole American Memory has provided a matchless and innovative set of resources for embedding the past in the present.[22]

Although the NDLP is the nine hundred-pound gorilla of digital library projects, it is not alone. The Making of America (MOA) Project, currently centered at Cornell University and the University of Michigan and funded by the Mellon Foundation, seeks to put online materials on the history of the United States, especially journals and books published in the second half of the nineteenth century. So far it offers only three (*Harper's New Monthly Magazine*, 1889–1896; *Manufacturer & Builder*, 1869–1894; and *Scientific American*, 1846–1850), but project representatives promise to put up 1.5 million pages relatively soon. Disappointingly, the user gets these journals only as pictures of pages, which makes them slow to download over a modem and prevents searching by word—a key advantage of most online texts.[23]

Many other libraries and archives are busy scanning documents and books for online presentation. For example, the Gilder Lehrman Institute of American History has put up 46 slave narratives. Wake Forest University has posted its Confederate Broadside Poetry Collection, which includes more than 250 poems written by Southerners and Confederate sympathizers during the Civil War. With support from Ameritech, the Library of Congress is offering grants to libraries, archives, and museums to enable them to create digital materials that will complement and enhance the library's American Memory collections.[24]

Creating Online Archives

The Web offers not just digital versions of existing archives but also entirely new archives designed specifically for the Web. Standing somewhere between a personal narrative and an archive, such an "invented archive" may highlight its subject in intriguing ways. Many historians have seen the pioneering Anti-Imperialism in the United States, 1898–1935, maintained by Jim Zwick.[25] A Ph.D. candidate at Syracuse University, Zwick began in 1994 by placing materials from his dissertation and related primary sources on the Internet; the collection has since grown to hundreds of documents, mostly primary sources. The University of Virginia's equally impressive but far more ambitious Valley of the Shadow site lets users explore two communities on either side of the Mason-Dixon line during the Civil War—although, at this point, only material for 1857–1861 has been placed on the Web.[26] Directed by Edward Ayers and based on a book idea of his, the project collects documents pertaining to Chambersburg, Pennsylvania, and Staunton, Virginia, allowing users to compare the two towns from the ground up. It includes the searchable contents of four newspapers, census data, records of the Union and Confederate armies, personal diaries, and miscellaneous documents, including maps and photographs. Aside from the Library of Congress, no single site offers a more eclectic and complete range of material online.

Valley of the Shadow is perhaps best understood as an ambitious new social history book on the Civil War, with all its primary sources available online. It allows students to construct their own narratives of life in both towns in the years before the war, but it seems to encourage narratives that follow the framework of Ayers's planned book. More than other sites, it seems clearly designed for teaching in a specific framework rather than for general reference.

The census data allow users to search for a specific name, occupation, gender, or level of wealth in Franklin County, Pennsylvania, and Augusta County, Virginia. A search for the last name "Stevens" in Franklin County produced 4 entries; a search for the occupation "seamstress" turned up 111 names, including those of two men. In an outstanding introduction to social history methods, students can investigate general categories in Franklin or Augusta counties or track individuals through the years before and during the war—including any service in the army and any mention of them or their families in local newspapers.

The project summary describes the newspapers as the "meat" of the archive, and you can search them extensively. If you search for a word from a particular newspaper, you are first taken to a summary of the relevant articles; you then have the option of downloading an image of a single newspaper page, which looks very much like a page of microfilm. The steps to reach this page may seem a bit awkward but the richness of the sources available—the full text of four newspapers—more than compensates. Recognizing the awkwardness of the searches and the long download times, the project staff has begun to transcribe hundreds of the articles. These can now be searched through familiar topics (for example, African Americans, Churches/Religious Activities, and Military/War News). Clicking on "Women" in the section on newspapers from Augusta County, Virginia, produces a series of summaries, each of which links to the full text of the article. A feature of the page encourages comparisons; students can scroll to the bottom of the Augusta "women's page" and toggle to articles on women in Franklin County.

Students can search other topics. From the table of contents, clicking on "Lincoln's Election" brings up a page with the voting breakdown and some political summary. It contains links to responses in local and national newspapers and to their treatment of Abraham Lincoln before and during the election. Clicking on "Race Relations" brings up links to articles about race relations in both regions.

The introductory page allows users to read the project staff's "first take" on the story to help orient students toward the best ways of seeing the archive, and it urges them to "construct their own narratives, coming up with ideas that eluded us." Students *can* use the data to come up with new narratives about individuals or institutions in the region. It seems harder to come up with alternative "metanarratives" of the war and its cause and meaning. The site appears more open-ended than it is—it tends to encourage users to follow the narrative framework and assumptions established by the comparative methodology. A more flexible structure might encourage other forms of comparison—of elite with non-elite, political with social, one political interest with another, or one race or gender position with another. Although it offers useful links to related Web sites where other resources can be explored, Valley of the Shadow is less a superficially neutral topical archive, like the Library of Congress's Conservation collection, and more a set of data organized around a specific historical thesis and methodology. But this organizational scheme has the powerful advantage that, as in a good book, there is drama and an easy-to-understand framework embodied in the story of two communities grappling with the war.

The clarity and drama of the story line make Valley of the Shadow particularly useful in the classroom. The site provides excellent examples of how teachers have used the site in their courses. It is probably the most sophisticated historical site on the Web, though it suffers from an inelegant and sometimes confusing interface. For example, the first page offers users three somewhat enigmatic categories: "The Impending Crisis," "The Communities," and "Sources." We should note, however, that Valley of the Shadow is now undergoing a redesign aimed at improving its speed, accessibility, and appearance. As with other Web sites (and unlike a book or article), one exciting feature of Valley of the Shadow is that it is a work in progress rather than a fixed and final product. Valley of the Shadow will eventually be accessible through a commercial CD-ROM, which will link to the database on the Web and reduce download times. Even a quick perusal of this site should help answer critics who say that the Web cannot be used for serious historical teaching. Indeed, sites like Valley of the Shadow and American Memory demonstrate that, at the present time, the most powerful uses of the Web for professional historians are in teaching, rather than in researching, the past.

Presenting the Past Online: Museums, Commercial Sites, and Personal Histories

Projects such as NDLP and Valley of the Shadow use the Web as a resource—a place to go to find specific information. Are there ways that the Web improves on well-established methods of turning historical facts into narratives, such as books, films, and museum exhibitions? Some promising experiments have emerged in the past year.

The Web site for the Computer Museum Network showcases the interactive possibilities of an online museum. The visitor registers before entering the site, and the look of the site is (lightly) customized based on the visitor's age and background. It offers such interesting features as the ability to send messages to other visitors (although our own attempts to do so failed), to participate in group solving of an online puzzle, to contact staff members and offer feedback, to propose donations of artifacts to the museum, and even to send a personally designed postcard.

The historical content of the Computer Museum Network site does not match its polished professional design. The historical exhibits are organized around a computer time line covering 1945–1979, which focuses narrowly on the nuts and bolts of computing and traditional history of technology (a chronology of firsts). Important developments are ignored: 1945 and 1946 include the development of the ENIAC (Electronic Numerical Integrator and Computer, the first working and general-purpose electronic computer), although nothing is said on its larger context, but the display ignores the publication of Vannevar Bush's article "As We May Think" in *Atlantic Monthly*, which set out basic premises of hypertext around which the Web itself is organized.[27]

Another limitation is that the designers of the site decided to keep it self-enclosed. You are never offered the opportunity to explore related resources on the Web, presumably one of its great advantages as a globally linked medium. The visitor has no way of knowing that the University of Pennsylvania has an extensive online exhibit on the development of the ENIAC and that the Smithsonian Institution has posted online a long interview with J. Presper Eckert, its codeveloper. This effort to erect a wall around the site means that, at least for historical information and understanding, the Computer Museum Network offers little more than a good, illustrated encyclopedia article. In general, there seem to be two impulses—whether conscious or unconscious—at war on the Web. One considers a site merely a piece of a larger network of information that the site's creators do not control; at one extreme, some sites are simply links to other sites. The other strategy—particularly strong for some institutions and for commercial operations—is more proprietary and attempts to capture the attention of the Web browser for its particular site.[28]

The greatest virtue of the Computer Museum Network and most online exhibits is simply remote access; the virtual museum brings objects and images (albeit often in less than fully satisfactory formats) to the person who cannot actually visit them. For example, if you missed the fine exhibit on the Ashcan school mounted by the Smithsonian's National Museum of American Art at one of its stops, you can see a nicely designed online alternative. The paintings are better seen in person, but on the Web site they arrive with historical material from the catalog—and this museum never closes. The Museum of the City of New York, "a museum for a new century," has placed a few exhibits online, including an excellent exhibit on Currier and Ives with sample prints and explanatory text. On the West Coast, the Museum of the City of San Francisco's online exhibit about the 1906 earthquake includes dozens of oral histories along with newspaper clippings and the official reports of relief agencies and public utilities—an outstanding resource. Thus online exhibits can become hypertexts that allow visitors to explore topics that interest them in much greater depth—a virtue appreciated by curators, who are continually told by designers to limit the amount of text they put on the wall.[29]

Aspects of Web technology hint that the "virtual museum" may eventually do some things better than its real counterparts. Interactive possibilities are tentatively explored by the Computer Museum (you may talk to visitors from Tokyo as well as to members of your own family). Another possibility lies in re-creating experiences or even sites that no longer exist. At a Web site devoted to Chetro Ketl Great Kiva (a subterranean sanctuary constructed by the Chaco Anasazi around a.d. 1000) developed by John Kanter, an anthropology student at the University of California, Santa Barbara, users with a properly configured browser can take a "virtual reality" tour of the kiva, descending a ladder into the smoky chamber.[30] Looking 360 degrees around a rendering of the kiva's interior, they can zoom in on particular objects while they listen to recordings of music by Plains Indians. A text accompaniment explains what they are seeing. Though not aimed specifically at historians, this site suggests the extraordinary ways the Web's multimedia potential might be used to re-create lost historical spaces.

A single (albeit energetic) graduate student produced the kiva site, with no budget to speak of. Neither a film nor an exhibition nor a static text, the site combines elements of all three. Will such nonprofit projects—from dedicated individuals or museums and historical societies relying on government and foundation support—be able to compete with those from private corporations?

To consider that question and to find out if the Web can improve on existing forms of popular history—films, books, and magazines—we point our Web browser to the HistoryNet, which bills itself as "the most extensive and content-rich site devoted to history on the Internet."[31] The core of the site is materials from the Web site's owner, "the History Group of Cowles Enthusiast Media," which publishes twelve historical magazines from *Aviation History* to *Civil War Times Illustrated* to *Wild West*. Not surprisingly given its origins and resource base, the site emphasizes topics of interest to history enthusiasts. On a typical week in November 1996, the six stories featured on the site's first page included Confederate generals J. E. B. Stuart and John Hunt Morgan, Captain Kidd's last voyage, and the blitz bombing of London. Enthusiasts who visit this site will probably find their preferences confirmed rather than challenged or broadened. The site's useful search engine yields twenty references to Gen. George McClellan but none to Eugene Debs or Marcus Garvey. There is nothing exciting or surprising about the HistoryNet's presentation of historical narratives. The stories, taken directly from the pages of the Cowles magazines, appear onscreen with a single illustration, a less attractive option than browsing the magazines.

Other parts of the site, however—a daily quiz and a useful "guide to exhibits and events"—make more use of the Web's interactive possibilities. Most interesting are the five online forums. They are not extremely active (fewer than 200 responses logged in the site's first six weeks), but the editors had started things off with thoughtful questions, such as "What drives your exceptional interest in the Civil War?" The questions provoked some revealing responses that would interest a professional historian who wants to know more about history enthusiasts or, better yet, to engage in discussion with them. Nevertheless, like many commercial sites, the HistoryNet is somewhat self-enclosed. The articles do not lead you to related materials elsewhere on the Web; even the guide to events and exhibits does not take you to pages maintained by the organizers of those exhibits.[32]

The financial basis of the HistoryNet is unclear; perhaps Cowles plans to include advertisements or to maintain the site as a useful advertisement for its own magazines and products from its online store, which sells everything from books to "Old Western Train" humidors. A more direct attempt to use the History Web to capture advertising dollars waits at the Discovery Channel Online. This Web site, which cost $10 million to set up and initially maintain, organizes its fare into six categories. History leads this group, and on one recent day users could find stories on the women pirates Anne Bonny and Mary Read and their brief careers with Calico Jack Rackham; Charles Lindbergh; and football broadcasting. The story on the female pirates includes short biographies of the major figures and sidebars on whether privateers were pirates and on the "pirate's code." Though it lacks historical citations, the site refers users to relevant books. The section on Lindbergh describes his fame and its eclipse in the late 1930s, when he appeared pro-German in his sympathies. Written by H. J. Fortunato, a "writer and business consultant," the story is lavishly illustrated and opinionated, a tasty historical "snack" rather than a serious work.[33]

Some Discovery Channel stories, on the other hand, have contained nuggets of useful and interesting information, obviously influenced by recent historiography. An earlier feature on "muscular Christians," for example, presents the familiar story of American religion's confrontation with modernity in colorful, informative ways. It conflates Father Charles Coughlin, Anthony Comstock, Carry Nation, and Father Divine—theologically suspect at the very least—but the organizational conception, spiced with audio and striking images, makes for entertaining and useful browsing. Another story, on the dust bowl, takes users hour by hour through a dust storm in Kansas in 1935. Richly and cleverly illustrated, it includes fifteen oral histories of the dust storms and their aftermath. Users can click to hear Thelma Warner remember how the dust "seemed to come right through the walls" or to hear Harley "Doc" Holladay tell a joke about farmers in hell. This page offers novel primary sources, framed with professional polish and skill.

Sites like this reinforce the criticism that well-funded corporate media could dominate the Web. One of the authors has established his own course-related Web page on the dust bowl, with no funding, little assistance, and no professional design or computer training. A comparison of the two sites reveals just how heavily the deck is stacked against academic historians and independent enthusiasts. Our site has few images, two audio clips, both from Woody Guthrie, a brief introduction, and a few excursions into other primary sources. The Discovery Channel feature on the dust bowl, written by Lori Ann Wark, does an excellent job of presenting the prevailing interpretation of the dust bowl's cause in a simple way, enlivened by oral histories and pictures. It includes bibliographic references to standard works and Web links to other related sites. Like the site as a whole, it reflects the impact of recent historiography in subject matter and treatment. The combination of good history and good production values is unusual; most commercial sites have only the latter. Is the dust bowl feature better than a book or a TV show? Yes and no—it has less substance than a typical book, but through oral histories it conveys more of the flavor of the people and their culture. Less passively entertaining than TV, the site offers twenty-four-hour access, and computer-savvy students could easily incorporate quotations from the material it offers in papers or other forms of presentation.[34]

Will heavily funded, professionally designed sites like this push politically controversial and independent history off the Web? It remains to be seen—the Discovery Channel has yet to recoup its investment. In the first half of 1996, the Discovery site generated only $573,000 in advertising revenue. But sites like this and the HistoryNet command huge resources and threaten to outshine the amateur and educational sites lovingly crafted largely by volunteer labor. Ironically, Cowles, as the champion of the history enthusiast on the Web, could crowd out the voice of the real amateur. The danger is the one that Herbert Schiller cites—that as the Web goes down the well-worn road of radio, TV, and cable, in which large "infotainment" conglomerates come to dominate the wires, "choice" becomes narrowly defined as the competition between two or three very similar "products." In this narrowcast vision of the Web, you will get history at the Discovery Channel, the History Channel, or the HistoryNet, but nowhere else.[35]

The potential for commercial versions of the past to crowd out nonprofit and amateur efforts on the Web comes not, as with TV, because there is limited "bandwidth" (only a certain number of channels) or because it is enormously expensive to own a television station. At least for the foreseeable future, everyone will be able to mount a personal Web page. But as the Web becomes more dominated by multimedia glitz, amateur and professional historians will find it increasingly difficult to compete with commercial operations. Where will they get the money to pay for permission to use a swing tune or a radio program from the 1930s in a site on New Deal culture? Who will provide the design and programming skills for flashy pages with interactive features? Self-made home pages will not disappear, but they might wither away in the face of competition from professionally developed and mounted sites. Like low-budget documentary filmmakers in an era of Hollywood blockbusters, they would find it hard to get anyone's notice and sustained attention.

Start-up costs for mounting sophisticated Web pages may become much more formidable. Most universities now provide faculty and students with access to equipment that is close to the technical cutting edge, which makes possible innovative sites like the Chetro Ketl Great Kiva. But there is no guarantee that such support will continue. Just as independent filmmakers now lack the equipment that offers the best special effects, independent Web page makers may come to lack access to the best computers, best software, best programmers, or fastest Web connections.

At the moment, however, literally hundreds of homemade sites densely populate the Web. Many explore topics that have traditionally attracted history amateurs and enthusiasts—the Civil War, military history, the West, and collectibles. The scale of this amateur effort can be glimpsed on the American Civil War Homepage, itself a volunteer effort, which contains links to almost 300 sites related to the Civil War (including 41 just for reenactors)—many created by unpaid enthusiasts. Larry Stevens, a phone company worker from Newark, Ohio, maintains 10 Civil War Web sites, mostly concerning Ohio in the Civil War. The sites combined his two hobbies of history and computers, and, he explains, he "decided to carve a niche into the net before the big boys aka Ohio Historical Society, Ohio State University, etc., entered the field."[36]

There are also many historical pages on which amateur historians treat such diverse subjects as the pioneering cartoonist and animator Winsor Mc-Cay, the landscape photographer William Henry Jackson, the history and future of money, and the magician Harry Houdini.[37] In these cases, people have taken materials gathered through personal passions and made them available to a much broader public. Sometimes these sites suffer from outdated links, since unpaid enthusiasts may lack the resources for active maintenance of their sites, which can become, in the new catchphrase, "cobwebs."

Sites started for one purpose can take on a larger life. John Yu, now a Microsoft employee, started his Web site on Japanese American internment for an undergraduate history class at the Massachusetts Institute of Technology (MIT), but it has now grown into a large site with primary documents, pictures, and a clear time line. Yu's Web site is less informative than most good secondary sources on the internment, but it differentiates itself from its print

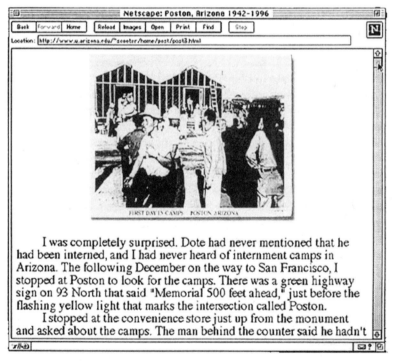

FIGURE 9.3 Scott Hopkins's aesthetically impressive Web site on the Japanese American internment camps in Poston, Arizona explores the intersection of past and present from a layman's perspective, combining personal narrative, oral interviews, and graphic reinterpretations of historical artifacts.

cousins by using the hypertextual features of the Web. You can click on most of the names and key terms in his documents and get a glossary or a biographical entry. More important (and unlike some other sites), it opens itself to a literal web of other resources on the internment and Japanese Americans generally. Thus, visitors to Yu's Japanese American Internment site might quickly find themselves stopping by the Unofficial Nikkei Home Page, created by a junior high school social studies teacher in Washington State; an online exhibit on the internment of a Japanese American from the Santa Clara Valley organized by Erin Kamura and the Japanese-American Resource Center in San Jose, California; a site on the Manzanar internment camp developed by two California high school students; or the University of Arizona's online exhibit, War Relocation Authority Camps in Arizona, 1942–1946. If you are lucky, you might wander into Scott Hopkins's "Web art site," which features a stunning presentation linking Poston, Arizona, camps and the game go. Hopkins uses photographs of the camps in the 1940s and the present, which he has beautifully transformed into color picture postcards, to reflect on the relationship of the past and the present.[38] Since little remains of many of the original internment camps, these virtual re-creations may be the most important historic sites connected to the internment experience.

Hopkins's site makes the intersection of personal interest and professional history vivid and intriguing. It shows us how dialogue about the past's meaning proceeds among professionals, enthusiasts, artists, educators, politicians, in casual hobbies and obsessive interests, from communities to university classrooms. More important, it comes to us in a format that stresses the multiple meanings of the past and the multiple sources of information to make sense of it. The Web clearly demonstrates that meaning emerges in dialogue and that culture has no stable center but rather proceeds from multiple "nodes." It also suggests the force of history in American life. Arguments that Americans are an "ahistorical people" or allow their history to languish collapse in the face of the range of historical material the Web contains. Like American society, the Web moves from slick, light commercial histories to stodgily responsible academic narratives, from irresponsible rant to thoughtful critique. The size and brute power of the Web—the ability to link quickly sites created in very different places and by very different people, to access distant material from your home or office, to search out specific words in unfamiliar contexts—makes it an extraordinary tool for making new *connections*, whether those are between the past and the present or between different concepts and bodies of knowledge.

The Web's very openness brings with it the threat that commercial operations will come to dominate and squeeze out the enthusiasts, cranks, and academics. In December 1996, the extraordinarily powerful Microsoft Corporation announced an overhaul that would make its Microsoft Network "look and feel a lot more like television."[39] Although we are loathe to underestimate the power of global infotainment conglomerates, which have taken control of other media (such as cable) once seen as conduits for democratizing cultural life, we do not think the future is predetermined. Like television stations, commercial Web sites use visual dazzle to hold your attention and limit "surfing." But unlike television, the Web allows alternative or contrarian viewpoints to flower, and it encourages users to compose their own narratives of the past. Academic historians, like other citizens, should insist on a role in this new "public space," should demand that it remain open and accessible to all, and should resist the tendency of television to wash political content toward the center. Universities, with their subsidized spaces for students, faculty, and staff, remain one of the best sources of experiment, of alternative viewpoint and serious content. The Web may not be the brave new world or the postmodern inferno, but it is an arena with which everyone concerned about the uses of the past in the present should be engaged.

Wizards, Bureaucrats, Warriors, and Hackers

Writing the History of the Internet

T AKE A LOOK AT THE STANDARD TEXTBOOKS on post–World War II America. You will search in vain through the index for references to the Internet or its predecessor, the AR-PANET; even mentions of "computers" are few and far between. The gap is hardly a unique fault of these authors; after all, before 1988, the *New York Times* mentioned the Internet only once—in a brief aside. Still, it is a fair guess that the textbooks of the next century will devote considerable attention to the Internet and the larger changes in information and communications technology that have emerged so dramatically in recent years.[1] Most historians will feel compelled to reckon with the Internet's emergence as a standard feature of everyday life.

How will that history be written? Four recent works offer some clues by addressing the history of the Internet from different perspectives (biographic, bureaucratic, ideological, and social) and considering different sources for the "creation" of the Internet—from inventive engineers and solid government bureaucrats to the broader social context of the Cold War or the 1960s. Although the Internet may be heralded as an entirely novel development, its historians have generally followed some well-worn paths in the history of technology. These conventional approaches are often illuminating, but the full story will only be told when we get a history that brings together biographical and institutional studies with a fully contextualized social and cultural history. The rise of the Net needs to be rooted in the 1960s—in both the "closed world" of the Cold War and the open and decentralized world of the antiwar movement and the counterculture. Understanding these dual origins enables us to better comprehend current controversies over whether the Internet will be "open" or "closed"—over whether the Net will foster democratic dialogue or centralized hierarchy, community or capitalism, or some mixture of both.

"Contextualist" approaches have long dominated academic studies of the history of technology, but narratives of "great men" of science and technology remain popular, deriving their power from widespread assumptions about new ideas emerging from particular men of genius as well as from the narrative appeal of biography.[2] The title of Katie Hafner and Matthew Lyon's well-written and extensively researched work of popular history, *Where Wizards Stay Up Late: The Origins of the Internet*, neatly inscribes the book's great man approach. So does the dust jacket, which promises "the fascinating story of a group of young computer whizzes . . . who . . . invented the most important communications medium since the telephone."[3]

Hafner and Lyon begin their tale of "origins" with Bolt Beranek and Newman (BBN), the computer consulting company that had the initial contract from the Advanced Research Projects Agency (ARPA) for what became known as the ARPANET. (Founded in 1957 in the post–Sputnik panic over Soviet technological prowess, ARPA, a Defense Department unit, supported research and development in technology, particularly military-oriented systems like ballistic missile defense.) The book's prologue describes a reunion of ARPANET's designers at BBN in 1994. This narrative choice and the centrality of BBN to the entire book owe a great deal to the study's origins in a suggestion from BBN, which opened its archives to Hafner and Lyon and even helped fund the project.[4]

Having started with the contractor, Hafner and Lyon explain the source of the contract with another story. As they tell it, Bob Taylor, the head of the ARPA office that dealt with computer research (known as the Information Processing Techniques Office), faced an "irksome" problem in the winter of 1966. The room next to Taylor's office housed three computer terminals, each connected to a mainframe running at a different site funded by ARPA. Since the different terminals reflected different computer systems, program languages, and operating systems, they required different login procedures and commands. "It became obvious," Taylor later remembered, "that we ought to find a way to connect all these different machines" and, thus, share extremely expensive computer equipment. "Great idea," his boss responded. "You've got a million dollars more in your budget right now. Go."[5]

After Taylor won funding for his project, he turned to "a shy, deep-thinking young computer scientist . . . named Larry Roberts" who was "blessed with incredible stamina" and "had a reputation for being something of a genius" to "oversee the design and construction of the network." In 1967, at a meeting in Ann Arbor, Wes Clark of Washington University came up with the crucial idea of making the network function by inserting a subnetwork of smaller computers between the host computers and the network lines—what later came to be called Interface Message Processors or IMPs. Riding to the airport in a cab, Clark told Roberts that only Frank Heart could build such a network at a reasonable cost. Heart too is a wizard: "intensely loyal" and "nurturing," he has "prodigious energy" and the ability to make "certain that jobs he signed up for really got done." And with his help, BBN, the Cambridge consulting company where he worked, snared the million-dollar contract to build the ARPANET. (When BBN won the contract for the Inter*face* Message Processors, Senator Edward Kennedy sent them a famous telegram congratulating them on the "ecumenism" of their planned work on the "Inter*faith* Message Processor.")[6]

But why begin with Taylor and BBN? Many popular narratives of the rise of the Internet start earlier and with a story that is more grounded in a particular historical context. A widely distributed "Brief History of the Internet" by science fiction writer Bruce Sterling opens: "Some thirty years ago, the RAND Corporation, America's foremost Cold War think-tank, faced a strange strategic problem. How could the US authorities successfully communicate after a nuclear war?" The solution, as Sterling explains it, emerged in 1964 from the Rand Corporation and particularly from engineer Paul Baran, who imagined a network with no central authority, which "would be designed from the get-go to transcend its own unreliability."[7] Unlike a centralized network in which destroying the central switching point brings down the entire structure, Baran theorized that a distributed network could sustain multiple hits and keep working through alternative channels. Crucial to Baran's distributed network was his second key innovation, using digital technology to break up messages into discrete pieces that could be sent individually and then reassembled at the end point—a feature that builds more reliability into the system, and makes more effective use of communications lines than telephone circuit-switching technology. (Telephone circuits set up a dedicated line between two people through which they send a continuous stream of words; if the participants turn silent for a minute, they continue to use the circuit. "Packet-switching networks" are much more efficient because the data are broken into smaller chunks, which can flow through multiple paths and also share the same lines with other pieces of data.) British physicist Donald Davies, who later developed some similar networking ideas, gave Baran's "message blocks" the name "packets"—a rubric that has stuck today and is embodied in the notion of "packet switching networks"—the core technology of the Internet.[8]

Starting with Baran instead of Taylor roots the Internet in the darkness of the Cold War rather than the bright idea of a clever engineer and emphasizes surviving (or fighting) nuclear war rather than sharing computer resources. Baran's work, he later told an interviewer, "was done in response to the most dangerous situation that ever existed." Like his contemporary at Rand, Herman Kahn (the model for "Dr. Strangelove" in the Cold War satire that appeared the same year as Baran's report), Baran thought the unthinkable—how to carry on after a nuclear apocalypse. "If war does not mean the end of earth in a black-and-white manner," Baran wrote, "then it follows that we should do those things that make the shade of gray as light as possible: . . . to do all those things to permit the survivors of the holocaust to shuck their ashes and reconstruct their economy swiftly."[9]

Hafner and Lyon do not ignore Baran, but they downplay his significance as part of deemphasizing the military origins of the Net even while they make clear that Baran's ideas were crucial in the development of the ARPANET. They credit Baran with putting in some of the Internet's "blocks" and "stones" but not with being its "architect." Roberts himself later put Baran more in the center of things, noting that when he read Baran's reports in 1967: "suddenly I learned how to route packets. So we talked to Paul and used all of his concepts and put together the [ARPANET] proposal."[10] But the real point for Hafner and Lyon is about intentions, not credit; the ARPANET, they insist, "embodied the most peaceful intentions to link computers at scientific laboratories across the country so that researchers might share computer resources . . . ARPANET and its progeny, the Internet, had nothing to do with supporting or surviving war—never did."[11]

Starting with Taylor's effort to connect disparate computers, Hafner and Lyon weave a lively tale of the origins of the Internet. But their biographical focus slights the technical and intellectual (as well as the military) roots of the ARPANET experiment: the influence, for example, of work on time-sharing computers (machines set up so that they can be used at the same time by multiple users), small scale computer networking projects, and the larger vision of giving people access to the world's knowledge—a heritage that runs from Diderot's *Encyclopédie* to H. G. Wells's "world brain" to Vannevar Bush's "memex" to J. C. R. Licklider's "libraries of the future."[12] By deemphasizing the social and political contexts in which the Net was built, Hafner and Lyon tell a story that most engineers would like—a tale of adventurous young men motivated by technical curiosity and largely unaffected by larger ideological currents or even narrower motives of self-advancement or economic enrichment.

Given their interest in the engineers and in BBN, Hafner and Lyon devote most of their book to a fast-paced narrative of the design and building of the system. They excel at explicating technical matters for a nontechnical audience. But their coverage trails off after they describe the first public demonstration of the ARPANET at the International Conference on Computer Communication in Washington in October 1972. Although that event established the feasibility of packet switching, success at this point was limited. No one had really figured out what the network was good for; as late as the fall of 1971, network traffic was barely 2 percent of what it could potentially handle; it was, as Hafner and Lyon nicely put it, "like a highway system without cars."[13]

The biographic, great man model stretches Hafner and Lyon's literary talents, in part because the Internet lacks a central founding figure—a Thomas Edison or a Samuel F.B. Morse. It resulted more from bureaucratic teams than inspired individuals. Bureaucracy, however, rarely makes for lively reading. A bureaucrat's story unfolds with great care and mastery, though little excitement, in *Transforming Computer Technology: Information Processing for the Pentagon, 1962–1986* by Arthur L. Norberg and Judy E. O'Neill. Just as funding, in part, explains Hafner and Lyon's focus on BBN, so too does funding explain Norberg and O'Neill's organizational focus. The book originated from a Defense Department contract to study the Information Processing Techniques Office (IPTO), with the original idea coming from the office's last director.[14] That support made possible an important set of forty-five interviews, which are extensively used in this book and also in a number of other works on the development of computing, including Hafner and Lyon's book.

Norberg and O'Neill consider not just ARPANET but all ARPA computer funding between 1962 and 1986, including for time-sharing, graphics, and artificial intelligence as well as networking. Although their book is scholarly in tone and in its extensive research and documentation, they champion their subjects just as Hafner and Lyon do. Throughout, the authors celebrate IPTO's "achievements," "contributions," "accomplishments," and "successes." The book also has its heroes—the bureaucrats who made everything happen. The authors devote one of the book's six chapters to describing and praising IPTO's "lean management structure." The agency's "technical accomplishments," they write, "were shaped as much by IPT office management as they were by researchers' intentions."[15]

By spotlighting ARPA, Norberg and O'Neill emphasize what Hafner and Lyon sometimes obscure—the close connection of all ARPA computer funding to military concerns. Calling their concluding chapter "Serving the Department of Defense and Nation," they celebrate rather than downplay that link. They point out, for example, that ARPA only set up the IPTO in 1962 in response to pressure from the Kennedy administration for improved military command and control systems.[16] Computers, it was widely believed, would make it possible to "control greater amounts of information and to present it in more effective ways to aid decision making." Whereas Hafner and Lyon describe IPTO's first director, J. C. R. Licklider, as pushing it toward basic research, Norberg and O'Neill quote him telling another military official that ARPA should only fund research that offers "a good prospect of solving problems that are of interest to the Department of Defense."[17] Such sentiments were hardly surprising from a man who went to work in the Pentagon the same month as the United States and the Soviet Union teetered on the brink of nuclear war over missiles in Cuba.

Norberg and O'Neill also provide a more complete and complex portrait of the Internet's ties to military concerns. They agree with Hafner and Lyon that Taylor's "perceived need to share resources" sparked his initial decision to seek funding for the ARPANET. But they also show that networking experiments grew out of IPTO's fundamental concern with using computers to improve military command and control. Norberg and O'Neill further argue that the military origins of the ARPANET made it successful. While "incentives for networking were lacking in the [computing] community," they "did exist in DOD [Department of Defense], where there was a need to reduce the high cost of software development, improve communications among military units while increasing computer use, [and] further develop command and control systems."[18]

In any case, to focus on the particular "originary" moment of Taylor's search for initial funding is to underplay the Internet's multiple origins. By 1972, ARPA had shown the feasibility of packet switching, but it had only created a limited and lightly used network, which also operated in a changed political climate. Starting in the late 1960s, White House and Congressional pressure forced ARPA to tie its funding much more closely to military needs.[19] In response to those mandates, ARPA sought to apply directly what it had learned about packet switching to military applications, particularly through packet radio networks and packet satellites. As the additional networks as well as some early commercial networks emerged, Bob Kahn, an engineer who had moved from BBN to ARPA in 1972, and others realized that they had now replicated the problem that had vexed Taylor back in 1966: how do you connect incompatible networks—rather than just computers—to each other? (Kahn, interestingly, had a direct connection to one of the Internet's key alternate origins; it was his cousin Herman Kahn's works on thermonuclear war that had provided the Cold War context for Baran's work on packet switching.)[20]

Out of this military-driven dilemma of "inter-networking" came both the concept and the name of the Internet. Kahn launched the "Internetting Project" to make it possible for "a computer that's on a satellite net and a computer on a radio net and computer on the ARPANET to communicate uniformly with each other without realizing what's going on in between."[21] In collaboration with Vinton Cerf, Kahn developed in 1974 a new and more independent packet-switching protocol—at first called Transmission Control Protocol or TCP and later TCP/IP, with IP standing for "Internet Protocol"—that would serve as a kind of lingua franca for this new Internet. It remains in use today. Not only did military funding and necessity create this standard but also the Defense Department's adoption of the protocol in 1980 for its own operations gave it a crucial boost. Equally important (and surprising given the context) was the Defense Department's public release of TCP/IP—in effect, this normally closed and secretive agency fostered a remarkably open (and hence free) standard of communication.[22]

But the ultimate triumph of TCP/IP was also—as Janet Abbate's informative dissertation makes clear—a matter of international politics and commerce. European telecommunication companies, publicly controlled, pushed an alternative standard (x.25) that would be more compatible with their operations. A key American weapon in the "protocol wars" was Defense Department support, which grew at least in part out of the explicit design of those standards for the military. As a result, TCP/IP boosters could, as Peter Salus notes in *Casting the Net*, persuade "the military brass that the ARPANET protocols were reliable, available, and survivable."[23] The victory of TCP/IP is not unconnected to why the United States still dominates the Internet.

Norberg and O'Neill provide a thorough institutional study but offer only passing references to the larger political and economic context. They acknowledge that the "political circumstances in the world of the past three decades led the Department of Defense to demand new developments in computing that would help to increase the sophistication and speed of new military systems," but add that "we will not discuss it in this study."[24] This lack of context also contributes to their largely uncritical view of ARPA's military mission. Despite the repeated references to military "benefits" and uses of the computer technology that ARPA funded, Norberg and O'Neill never discuss the actual use of computers on the battlefields of the Vietnam War, which was fought precisely during the heyday of ARPA funding of computer projects.

Although Paul Edwards's *The Closed World: Computers and the Politics of Discourse in Cold War America* does not focus specifically on the Internet, it still shares many topics and sources with the Norberg and O'Neill and Hafner and Lyon books. Nevertheless, it is also their mirror opposite: whereas Norberg and O'Neill as well as Hafner and Lyon eschew context, Edwards places his story squarely within the narrative of the Cold War and emphasizes the world outside the laboratory; whereas Norberg and O'Neill celebrate (and Hafner and Lyon deny) the marriage of defense and computers, Edwards paints a forbidding portrait of their union; whereas Norberg and O'Neill and Hafner and Lyon provide straightforward (and easy to follow) institutional or biographical histories, Edwards, as a student of Donna Haraway and a graduate of the History of Consciousness program at Santa Cruz, draws on and contributes to a large theoretical literature in cultural studies and structures his (sometimes confusing) account more as "collage than linear narrative." Edwards departs most sharply from other works in his abandonment of the trope of "progress" that often marks writing about the history of technology.[25]

The richness and complexity of Edwards's sometimes brilliant account make it difficult to summarize briefly.[26] Edwards contends that the digital computer is both cause and effect of what he calls the Cold War's "closed-world discourse," which he defines as "the language, technologies, and practices that together supported the visions of centrally controlled, automated global power at the heart of American Cold War politics." "Computers," he writes, "created the technological possibility of the Cold War and shaped its political atmosphere." And, in turn, "the Cold War shaped computer technology." Cold War politics "became embedded in the machines," including their "technical design," and the "machines helped make possible its politics." In this way, Edwards goes beyond historians who argue for the "social construction" of technology and focus on how different social groups shape the development of technology. He emphasizes instead what he calls the *"technological construction of social worlds."* Computers in this analysis, heavily influenced by the work of Michel Foucault, become themselves a source of power and knowledge—or in Edwards's words, "a crucial infrastructural technology—a crucial Foucaultian support—for the Cold War closed-world discourse."[27]

That the Cold War, if not Cold War discourse, fostered the development of digital computers is relatively easy to show.[28] In 1950, for example, the federal government—overwhelmingly, its military agencies—provided 75 to 80 percent of computer development funds. Even when companies began funding their own research and development, they did so with the knowledge of a guaranteed military market. Such massive government support enabled American computer research to destroy foreign (mostly British) competition; the American hegemony in computer markets—routinely attributed to American free markets—rests on a solid base of government-subsidized military funding. "The computerization of society," writer Frank Rose aptly observes, "has essentially been a side effect of the computerization of war."[29]

Such facts are relatively well known (although sometimes ignored by ideologues who depict the computer industry as the exemplar of laissez-faire), but Edwards wants to make a deeper argument about the significance of military involvement in computer development. He rejects the idea that "military support for computer research was . . . benign or disinterested"—a view he attributes to historians who take "at face value the public postures of funding agencies and the reports of project leaders." (He could be talking directly about the Hafner and Lyon and Norberg and O'Neill books, but their work appeared either after or at the same time as his book.)[30] Rather, he argues, "practical military objectives guided technological development down particular channels, increased its speed, and helped shape the structure of the emerging computer industry." For example, he maintains that the shift from analog to digital computing was not the result of the innate technological superiority of the latter but of the digital approach's better correspondence with and support for the vision of centralized command and control of the closed-world discourse.[31] Unfortunately, Edwards never makes clear precisely how computing would look different today without defense funding under the shadow of the Cold War. Would we have analog computers on our desks—or none at all?

Indeed, Edwards is more interested in showing that computer technology helped create and develop the discourse of centralized command and control than in exploring how this vision actually shaped computer design. Computers, he writes, "helped create and sustain this discourse" by allowing the "practical construction of central real-time military control systems on a gigantic scale" and facilitating "the metaphorical understanding of world politics as a sort of a system subject to technological management."[32]

Much of this sounds and is rather abstract, but Edwards leavens the book's relentless abstractions with a series of rich case studies and anecdotes. We learn, for example, about U.S. Air Force Operation Igloo White. Run from the Infiltration Surveillance Center in Thailand (the largest building in Southeast Asia) and costing nearly $1 billion per year between 1967 and 1972, Igloo White sought to monitor all activity across the Ho Chi Minh Trail in southern Laos, including truck noises, body heat, and the scent of human urine. When the sensors ("shaped like twigs, jungle plants, and animal droppings") picked up signals, they appeared magically on the display terminals as "a moving white 'worm' superimposed on a map grid." Then the computers would project the "worm's" movements and radio the coordinates to Phantom F-4 jets, whose computers would guide them to the precise map grid square; the computers back in Thailand controlled the release of the bombs. "The pilot," observes Edwards, "might do no more than sit and watch as the invisible jungle below suddenly exploded into flames." It was the perfect fantasy of the closed world of computerized and centralized command and control. In the apt words of one technician: "We wired the Ho Chi Minh Trail like a drugstore pinball machine, and we plug it in every night." But the "pinballs" were smarter than the players. The Vietcong fooled the sensors with taped truck noises and bags of urine, which duly provoked massive air strikes on empty jungle corridors. These airstrikes were then claimed as quantitative (and quantifiable) successes. A 1971 Senate report found that "truck kills claimed by the Air Force [in Igloo White] last year greatly exceeds the number of trucks believed by the Embassy to be in all of North Vietnam." Even if the exaggerated claims had been true, they could have only been scored as successes in a crazy world in which it would have cost $100,000 to destroy trucks and supplies worth a few thousand dollars.[33]

Igloo White, as Edwards shows, typified computerized Cold War military operations. He devotes a chapter to the Semi-Automatic Ground Environment (SAGE) computerized air defense system, which cost billions of dollars and was obsolete by the time it was fully operational in 1961. But in the irrational closed world of the Cold War, SAGE actually "worked," as Edwards argues. Computer scientists got to pursue their research; IBM Corporation built its dominance of the computer industry with the help of the massive SAGE contract. And on an ideological level, SAGE worked by "creating an impression of active defense that assuaged some of the helplessness of nuclear fear" and fostering the myth of centralized control and total defense.

Although Edwards has little to say directly about the ARPANET, it is difficult to read his book and then share Hafner and Lyon's or Norberg and O'Neill's view of the connection between the military and the rise of the Internet as accidental or benign. One of the sharpest differences between Edwards's account and the others is in the depiction of J. C. R. Licklider, who twice directed IPTO and whose famous 1960 paper on "man-machine symbiosis" helped shift computing from computation to communication. For both Hafner and Lyon and Norberg and O'Neill, Licklider is an almost sainted figure. "Everybody adored Licklider," Hafner and Lyon write. "His restless, versatile genius gave rise through the years to an eclectic cult of admirers." His "worldview," they write, "pivoted" on the idea "that technological progress would save humanity."[34]

In these other accounts, particularly Hafner and Lyon's, Licklider's concern with "man-machine" interaction appears as largely an intellectual problem. But Edwards maintains that it grew directly out of his World War II work in Harvard's Psycho-Acoustic Lab, which sought to reduce "noise" in battlefield communications systems. Such military concerns continued to inform Licklider's work after the war. In his 1960 paper, for example, he explains the problem with batch processing (as opposed to real-time interactive computing) by writing: "Imagine trying . . . to direct a battle with the aid of a computer on such a schedule as this." Edwards thus depicts Licklider as tightly wedded to military goals, describing him as "deeply desir[ing] to contribute to new military technologies from his areas of expertise." Writing in 1978, Licklider expressed some frustration that the World-Wide Military Command and Control System's computers were not yet "interconnected by an electronic network" and used an operating system designed for "batch processing." He argued that "military command and control and military communications are prime network applications" and observed that "both interactive computing and networking had their origins in the SAGE system."[35] But regardless of Licklider's own views, the Defense Department would never have committed funds to projects like ARPANET without the belief that they would ultimately serve specific military objectives and larger Cold War goals.

Thus it becomes clear that computer systems were invented for the Cold War, which provided the justification for massive government spending, and were pushed in particular technological directions. But these same computer systems, in turn, helped to support the discourse of the Cold War; they sustained the fantasy of a closed world that was subject to technological control. Even before ARPANET, the first real computer network was developed by the SAGE project because "the massive integration of a centralized, continental defense control system" required "long-distance communication over telephone lines."[36]

If the Internet, like networking and computing, in general, was a "side effect of the computerization of war," did it also support that militarized and closed vision of the world? On the one hand, the notion of a network of interconnected computers—especially one that could survive nuclear attack—fostered the fantasy of centralized command and control that Edwards sees as crucial to closed-world discourse. Moreover, at least in Defense Department hands, the ARPANET was quite literally a "closed world" to which only a select number of ARPA-funded sites had access. But, on the other hand, Baran's distributed network—perhaps precisely because it responded to a *post*-nuclear war scenario—could also have nurtured a highly decentralized view of the world. Norberg and O'Neill report, for example, that Defense Department officials initially viewed the new network with suspicion because it would "make it easier for subordinates to send messages without the approval of commanding officers, possibly circumventing the military's chain of command."[37]

And in the 1960s there were plenty of reasons to worry about subversion of the chain of command and of military thinking, in general—a fact that

Edwards's closed-world analysis seems to ignore.[38] He provides an often perceptive analysis of some of the key Cold War–era films, for example. But he does not give enough weight to the way that films like *Dr. Strangelove* (1964) both popularized the closed-world discourse but also undercut it by showing the idea of controlling the nuclear world to be an absurd fantasy. Some leading scientists also came to have doubts. In December 1968, fifty senior faculty members at MIT—the center for the most important developments in computing as well as the country's biggest academic defense contractor—circulated a statement that started: "Misuse of scientific and technical knowledge presents a major threat to the existence of mankind. Through its actions in Vietnam our government has shaken our confidence in its ability to make wise and humane decisions." That declaration led directly to the founding of the Union of Concerned Scientists early the next year; the group particularly challenged the conventional wisdom on nuclear weapons and fostered debate over military funding of academic research.[39] At least some scientists were beginning to question closed-world visions, and, indirectly, Edwards's own work emerges out of that critical tradition.[40]

Those creating the ARPANET could hardly have been unaware of these protests. Just six months before the network's first successful connection in October 1969 between UCLA and the Stanford Research Institute (SRI), massive student protests focused on SRI, calling for an end to all classified, chemical warfare, and counterinsurgency research. On April 18, 1969, 8,000 students and faculty at Stanford voted to commend the protesters for "helping focus attention of the campus upon the nature of research being conducted at the University and SRI."[41] Antiwar protesters across the country repeatedly targeted closed or classified research.

In addition to those who frontally assaulted the closed-world vision of the Defense establishment, there were those who took a less direct but still subversive approach. ARPA money supported the "hackers" at MIT's Artificial Intelligence Lab, but some of their goals—the free sharing of information, for example—led to direct clashes. Richard Stallman, a systems programmer at the lab, carried on a guerilla war against the use of passwords on the system. The lack of security encouraged by Stallman and others caused nervousness at the Defense Department, which threatened to cut the computer off the ARPANET since anyone could walk into the lab and connect to the rest of the network.[42]

An even more important question about the connection between closed world discourse and the Internet is how the new global network operated in practice. Edwards shows that military systems like Igloo White and SAGE did not work as planned. What were the actual workings of the ARPANET and Internet? To the biographical, bureaucratic, and ideological histories of the Internet, we need to add a social and cultural history.

Michael and Ronda Hauben's *Netizens: On the History and Impact of Usenet and the Internet* offers a strikingly different historical narrative of the Internet—one that insists that the real story is not of the "wizards" who built the Internet but of the "Netizens" who figured out what it was "really" for and popularized it. In their populist account, ordinary users who realized that it offered a marvelous medium for democratic and interactive communication created the soul of the new network from the bottom up. And while their book is sometimes repetitive and poorly written, it offers an interpretive perspective that should be central to any future Net history.[43]

The Haubens see the bottom-up origins of the Internet in "Usenet," the international computer newsgroup network that has more recently been overshadowed by the World Wide Web but still has a substantial presence on the Internet—more than 30,000 different newsgroups covering everything from alien visitations (alt.alien.research) to Zoroastrianism (alt.religion .zoroastrianism). In 1979, two Duke University graduate students, Tom Truscott and Jim Ellis, working with other students at nearby schools, developed some simple programs through which computers using the popular Unix operating system could call each other and exchange files. In effect, the system made possible an online newsletter that would be continuously updated. Those with access to any of the connected computers could read the news postings and add their own comments with the knowledge that they would be quickly read by everyone else; the same program allowed e-mail to be sent between the Unix computers connected by phone modems.

The graduate students consciously saw themselves as offering a networking alternative to the ARPANET, then still limited for reasons of cost and security to Defense Department-funded sites.[44] Several months later, they described Usenet as trying to "give every Unix system the opportunity to join and benefit from a computer network (a poor man's ARPANET, if you will)." Another of the graduate students, Stephen Daniel, later recalled that they had "little idea of what was really going on on the ARPANET, but we knew we were excluded."[45] The students' insurgent computer network grew with startling speed: from the initial three sites to 150 two years later, then jumping to 5,000 by 1987. In 1988, Usenet connected 11,000 sites, and participants posted about 1,800 different articles each day. Usenet grew along with the runaway popularity of Unix, which became the standard operating system for the 1980s. A crucial breakthrough had come in 1981 after Usenet gained a tenuous one-way connection from the ARPANET (linked between different computers at the University of California, Berkeley). When graduate student Mark Horton established this gateway, he pierced what some disgruntled Usenet participants described as the "iron curtain" surrounding ARPA-NET.[46] Barriers fell further two years later when the Defense Department segmented off its military communications into MILNET, which made it less nervous about what traveled over the ARPANET.

The runaway growth of Usenet as a forum for conversation and communication was paralleled by the earlier discovery of e-mail as the most popular use for ARPANET. In 1972, BBN engineer Ray Tomlinson, working on his own, developed a program for sending mail messages across the ARPANET. By the following year, three-quarters of network traffic was devoted to e-mail. Almost overnight, the empty highway found its cars; to this day, e-mail remains the most popular use of the Internet.[47] As with Usenet, e-mail had come from "below," from computer users, who wanted to communicate with other computer users, rather than from ARPA directives from above. And as with Usenet, the technology had emerged from someone "hacking" around, rather than carrying out an official plan.

Much of the Haubens's book is devoted to a somewhat hyperbolic celebration of Usenet and other computer networks as a democratic and "uncensored forum for debate" that is the "successor to other people's presses, such as broadsides at the time of the American Revolution and the penny presses in England." They argue that the Internet has created a new kind of citizen, the "Netizens," whom they define as "people who decide to devote time and effort into making the Net, this new part of the world, a better place"—"a regenerative and vibrant community and resource."[48] The Haubens see the democratic nature of the network growing out of its grassroots source in the people who created Usenet.

In addition to emphasizing this later moment of creation for the Internet and locating its paternity in the person of some Duke graduate students, the Haubens also give a more democratic and grassroots spin to the earlier history of ARPANET. In particular, they stress a moment in the development of ARPANET that others have described but not necessarily in the same populist tones. This came early in 1969 when BBN convened a "Network Working Group" to devise the protocols for the new network. Steve Crocker, a bearded young UCLA graduate student, agreed to write up notes from the meetings. Crocker framed his notes to emphasize that "anyone could say anything and that nothing was official." He labeled them "Request for Comments" and this ongoing series of "RFCs" (distributed ultimately through the medium of the network) became the way that Internet standards have evolved to this day.[49]

The Haubens, not surprisingly, celebrate the philosophy behind the RFCs as representing "unprecedented openness" that fostered the "amazing and democratic" achievement of the Net and its "cooperative culture." They also remind us that the decision to evolve technical standards in such an open-handed way came at a particular moment in time—the 1960s. "The open environment needed to develop new technologies," they write, "is consistent with the cry for more democracy that students and others raised throughout the world during the 1960s." Not surprisingly, the builders of the ARPANET were well aware of this context. Writing in 1987 on "The Origins of RFCs," Crocker recalls that "the procurement of the ARPANET was initiated in the summer of 1968—Remember Vietnam, flower children, etc?"[50] By placing the rise of the Internet within the 1960s-as-counterculture and the 1960s of the antiwar movement, Crocker and the Haubens suggest an alternative contextual frame to that emphasized by Edwards, who puts the rise of digital computing (and implicitly the Internet) solely within the Establishment 1960s of the Vietnam War and the Cold War.

Both contexts are, of course, important and suggest how we might revise Edwards's analysis to see the Internet as shaped both by the "closed world" discourse of the Cold War and by the "open world" discourse of the antiwar movement and the counterculture. Such an analysis would also incorporate the entertaining and revealing story Steve Levy tells in *Hackers: Heroes of the Computer Revolution*. Levy discerns among the hackers of the 1960s and 1970s (whom he defines as "those computer programmers and designers who regard computing as the most important thing in the world") a "philosophy of sharing, openness, decentralization, and getting your hands on machines at any cost—to improve the machines, and to improve the world." Although this "hacker ethic" was not simply the technological side of the counterculture and the antiwar movement, it drew from some of the same sources. "All over the Bay Area," Levy writes of the early 1970s, "the engineers and programmers who loved computers and had become politicized during the antiwar movement were thinking of combining the two activities." In 1972, for example, Bob Albrecht launched a tabloid called *People's Computer Company* (inspired by Janis Joplin's group, Big Brother and the Holding Company), which proclaimed on the cover of its first issue: "Computers are mostly used against people instead of for people. Used to control people instead of to FREE them. Time to change all that—We need a . . . People's Computer Company." Among the frequent visitors to the paper's potluck dinners was Ted Nelson, the author of the self-published manifesto of counterculture computing: *Computer Lib.*[51]

Berkeley's Community Memory project similarly merged the impulses of the radical 1960s with the hacker ethic by setting up a time-shared mainframe computer on the second floor of a record store and opening it to free, public use as a kind of combined electronic version of a public library, coffeehouse, urban park, game arcade, and post office. Community Memory embodied, as Levy says, the effort to take "the Hacker Ethic to the streets" and to allow people to use computer technology "as guerilla warfare for people *against* bureaucracies." Not coincidentally, some aspects of Community Memory—the decentralization and the free sharing of information— sound like the Internet. And Levy argues that the ARPANET "was very much influenced by the Hacker Ethic, in that among its values was the belief that systems should be decentralized, encourage exploration, and urge a free flow of information."[52]

Among the founders of Community Memory was Lee Felsenstein, a red diaper baby (son of a district organizer for the Philadelphia Communist Party) who had worked as an audio technician for the Free Speech Movement and spent the 1960s moving between seemingly contradictory existences as engineer and political activist. He embodied the two key groups that Martin Campbell-Kelly and William Aspray identify as the vanguard for the personal computer revolution of the early 1970s—computer hobbyists who emerged out of the world of radio and electronics aficionados and loved the idea of building their own equipment and, second, computer liberationists who emerged out of the New Left and the counterculture and loved the idea of bringing computers to the people. In the 1970s, Felsenstein became the moderator of the famous "Homebrew Computer Club," where computer hobbyists and computer liberationists came together to create the first PCs. (When Felsenstein made a big score himself by designing the Osborne personal computer, he plowed the money into Community Memory.) Activist and counterculturist hackers like Felsenstein, in effect, tried to turn the closed-world discourse on its head and make the personal computer and community networks into "supports" (to use Edwards's term) for a discourse of freedom, decentralization, democracy, and liberation.[53]

Some of the computer developments of the late 1960s and the 1970s, while less directly shaped by radical politics or the counterculture, still bear the imprint of the period. Ken Thompson and Dennis M. Ritchie, the bearded and longhaired Bell Labs' programmers who, in 1969, developed Unix, the operating system behind Usenet, later described themselves as seeking "a system around which a fellowship could form." As Campbell-Kelly and Aspray point out, "Unix was well placed to take advantage of a mood swing in computer usage in the early 1970s caused by a growing exasperation with large, centralized mainframe computers."[54] Protests in the 1960s had featured students wearing punch cards around their necks with the slogan "Do Not Fold, Bend, Mutilate or Spindle," but the hostility to the large mainframe computers and centralized batch processing extended beyond radical students to computer scientists and computer users who increasingly favored decentralized smaller computers, often running Unix.[55] Not coincidentally, Unix-style operating systems, not dependent on proprietary hardware and software standards, have become known among computer scientists as "open systems."

Still, it would be a mistake to collapse the story of computers and the Internet into the story of the radical sixties, as the Haubens do sometimes. When MIT went on "strike" on March 4, 1969, most students and faculty spent the day, as usual, in their labs and classes.[56] Moreover, many radicals wanted to smash technology rather than liberate it. In 1962, the Port Huron statement had lyrically celebrated the potential of science to "constructively transform the conditions of life through-out the United States and the world," but in 1964 Mario Savio, the son of a machinist, had spoken eloquently of the need to "put your bodies upon the gears and upon the wheels" to stop "the machine." And by the late 1960s, many counterculture adherents headed for rural communes.[57] To make the case for the impact of 1960s radicalism on the rise of networking requires a more precise social and political history. We need to know more about the graduate students who crafted the first "Requests for Comments." Some of them may have had beards, but most were also willing to take Defense Department funding, which their more radical counterparts would have eschewed. Such a wider social history would also probably help us see that the Internet and Usenet originated in a "community" but also a very specific kind of community—young graduate students and faculty in computer science and related fields. When those young engineers and scientists turned ARPANET into a mail system rather than a medium for sharing computer resources and formulated Usenet, they were participating in a "quest for community"—but the most important component of that community was technical knowledge rather than sixties-style politics and culture.

To be sure, there were signs of the 1960s on the early networks: drug deals and antiwar messages, for example, flowed through the ARPANET.[58] But the largest amount of traffic was initially about technical matters; the very first e-mail discussion group (MsgGroup), launched in June 1975, was about e-mail itself—participants argued heatedly about such fascinating topics as the proper format for e-mail headers. The first invitation to participate in Usenet promised discussions of "bug fixes, trouble reports, and general cries for help."[59]

As late as 1982, most ARPANET and Usenet discussion groups still focused on technical matters. Most other group discourse reflected the leisure pursuits of young male engineers and computer scientists—science fiction, football, ham radios, cars, chess, and bridge.[60] Only a few groups considered more broadly political topics like alternate energy production. While the Haubens romanticize the early days of Usenet and ARPANET as the nesting ground for a broad democratic community, it was the creation of a rather more specific form of community. The "MsgGroup," explained a Carnegie Mellon graduate student in 1977, "is the closest that we have to a nationwide computer science community forum." And for computer science students who were at schools not privileged to have an ARPANET connection, Usenet was, as one of them explained, "our way of joining the Computer Science community and we made a deliberate attempt to extend it to other not-well-endowed members of the community."[61]

Indeed, the rapid growth of computer science as an academic discipline in the 1960s and 1970s paralleled and fostered the rapid growth of the Net. In 1962, Purdue and Stanford universities set up the country's first two computer science departments; by 1979, there were about 120. That only 15 of these universities had ARPANET connections fostered the sense of exclusion that led Truscott and Ellis and other graduate students to create Usenet. Back in 1974, the National Science Foundation had proposed a network for academic computer scientists that would "offer advanced communication, collaboration, and the sharing of resources among geographically separated or isolated researchers."[62] In the early 1980s, that network emerged as CSNET, and, by the mid-1980s, it connected almost all U.S. universities' computer science departments. CSNET had connections into ARPANET, and it became one of several different networks (for example, BITNET) that would later be combined into the Internet.

While this quest for professional (and male) community may have lacked the political edge of 1960s radicalism, it drew on some of the remnants of a sixties-style ethos, which was still very much alive at universities in the 1970s. Even something as seemingly self-evident as e-mail was propelled by winds of change blowing from the 1960s. As Ian Hardy points out in his study of the emergence of e-mail, the medium's "disdain for false formality, its distrust of traditional hierarchy, its time-selfishness, speed, and certainly its ironic juxtaposition of impersonality and emotional directness" represented a "new culture of interaction" that might not have been so readily possible without what Kenneth Cmiel calls the "informalization" of culture that the 1960s brought.[63] In general, then, many of the "open" qualities of the Internet can be seen as rooted, at least in part, in impulses that came from the 1960s—the open process of creating standards through RFCs drew on challenges to hierarchy and commitments to candor; the rise of e-mail and newsgroups was influenced by a powerful quest for community as well as a growing informality in communication (both in habits of speech and in the rise of alternative newspapers); the interest in decentralized networks gained support from a distrust of large centralized structures, including centralized batch-processing computing and the desire to share information freely; and the rise of alternative networks like Usenet was supported by an effort to break down modes of exclusion. Ironically, while the Department of Defense had very different goals in mind—and often tried to implement them by, for example, restricting access to the ARPANET or to what it could be used for—its willingness to embrace the open technical standards embodied in TCP/IP inadvertently sparked the creation of a remarkably open system.

The apparent failure of the Cold War discourse to police its own boundaries suggests that what we think of as "sixties" hostility to conformity and hierarchy had much broader and deeper sources than just the counterculture, as Thomas Frank shows in his recent book on business and the counterculture, *The Conquest of Cool*. "The meaning of 'the sixties,'" he writes, "cannot be considered apart from the enthusiasm of ordinary, suburban Americans for cultural revolution."[64] A broader picture of the 1960s would, then, include computer science graduate students rejecting proprietary, hierarchically organized, batch-processing computer systems running on IBM mainframes as well as longhaired hippies smoking dope at Woodstock. Or maybe the closed world of the military and the open world of the hippies were not as separate as we sometimes think—at the heart of the military-industrial complex we might find beatnik Maynard G. Krebs with a math degree.[65]

In different ways, both Levy and the Haubens help us to see that the more profound challenge to this "open" vision of the Internet that was rooted (at least in part) in the 1960s came not from its heritage in the Defense Department but rather from an alternative, closed system—corporate capitalism. In 1975, after the first personal computer, the Altair, appeared on the cover of *Popular Electronics*, two teenagers, working from the plans, wrote a BASIC program for the new machine. But even before MITS, the Altair's manufacturer, officially released the program, bootleg copies circulated rapidly among computer enthusiasts imbued with the hacker ethic that "information wants to be free."[66] One of the teenagers, whose name was Bill Gates (the other was Paul Allen), wrote an angry "Open Letter to Hobbyists" arguing that people who wrote software ought to get paid. Gates's letter augured a new world in which, Levy writes, "money was the means by which computer power was beginning to spread."[67] Information could not remain free when people were paying large sums in cash.

For the Net, the transition from public or open to private and proprietary started around the same time and also quickly got entangled in questions of "ownership." In 1972, ARPA announced that it wanted to sell the network, but the major telecommunications corporations (including AT&T) showed little interest. Others more closely associated with the development of the new networks, however, saw money to be made. BBN, for example, set up its own subsidiary Telenet to provide commercial services and brought in none other than ARPA official Larry Roberts as the president of the new business. A dispute quickly ensued over whether BBN had to share the "source code" for the Interface Message Processors with their emerging competitors. In this case, government muscle forced BBN to make the code openly available, but it heralded a new era in which corporations would make huge sums off computer software initially developed at government expense.[68]

Telenet and some competitors drew directly on the open technologies developed by ARPANET. But some commercial firms took an opposite strategy. Large computer firms such as IBM and Digital Equipment developed proprietary networks—SNA and DECNET, for example—with the goal of keeping customers tied to their own hardware and software.[69] But ironically, the Defense Department's embrace of the "open standards" of the Internet doomed these efforts to failure. That failure did not, however, keep the Net from moving from a subsidized public good to an arena for profit making. In the 1980s, the National Science Foundation, which had taken control of the Internet from ARPA, moved to privatize it. Populists like the Haubens have bemoaned the transformation from public to private control and ownership, yet the change evoked remarkably little protest. In the 1980s, when most forms of publicly owned goods and services—from public schools and public housing to public parks—were in decline and an ideology of privatization and deregulation was in ascendance, it seemed like conventional wisdom to turn this public utility over to private ownership.

By the 1980s (and especially by the 1990s), moreover, many of the people who had celebrated the freedom and openness of networks and personal computers had also undergone a transformation that made them inclined to accept this privatization. The affection of many "Netizens" for free speech and freedom from control had also come to embrace a love for free markets. The liberationism of the many early computer and network enthusiasts had been transformed into libertarianism. "Technolibertarianism" became one of the central ideologies of the Internet. Many computer liberationists of the 1960s and 1970s now find themselves aligned with conservative free market prophets such as George Gilder and Alvin Toffler.[70] This may be less contradictory than it seems on the surface. As Mark Lilla has recently argued, "the cultural and Reagan revolutions took place within a single generation, and have proved to be complementary, not contradictory events." Americans, he writes, "see no contradiction in holding down day jobs in the unfettered global marketplace—the Reaganite dream, the left nightmare—and spending weekends immersed in a cultural universe shaped by the sixties."[71] In that sense, the Internet of the 1990s may be the perfect synthesis of the antihierarchical cultural revolution of the 1960s and the antistatist political revolution of the 1980s.

Yet this synthesis retains its own internal tensions and contradictions. While free marketeers today celebrate the Internet as the home of "people's capitalism," it also seems headed down the road to oligopoly. Three companies—the newly merged MCI WorldCom, Sprint, and Cable & Wireless— probably control three-quarters of the Internet backbone.[72] Web search companies, which are seen as the portals to the Internet, are busily gobbling each other up or being acquired by larger media conglomerates. Bill Gates's Microsoft Corporation has a pretty good chance of controlling not only all of the personal computers from which people access the Internet but also the browsers through which they read pages on the World Wide Web. And Intel Corporation is poised to be the manufacturer of choice for the chips at the heart of those computers.

Yet the road toward monopolization and centralized control is not preordained. The current antitrust cases against Microsoft and Intel—or, less plausibly, the revival of popular antimonopoly sentiments—might alter the corporate landscape. In general, the tendencies toward both open and closed systems that have shaped the Internet from its origins remain with us today. On the World Wide Web, we can find Web pages from every major corporation, but ordinary people still post their own pages with the same do-it-yourself enthusiasm as the members of the Homebrew Computer Club. (An astonishing 46 percent of Web users have created their own pages, according to one recent survey.[73]) Most Internet servers run Unix or Windows NT, but a surprising number (and 3 to 5 million people overall) use a freely distributed operating system called "Linux," which itself incorporates crucial components developed by the Free Software Foundation headed by Richard Stallman, the MIT hacker who violated ARPA security. And the most popular Web server software (Apache) and the most widely used programming language for Web sites (Perl) are also "freeware." (Finnish programmer Linus Torvalds first put together Linux in order to get access to Usenet, where he chronicled his progress in developing the software and sought help from other programmers.[74]) Commerce and advertising have infiltrated every corner of the Internet, but millions of people use the Internet to debate ideas or search for love in Usenet discussion groups, America Online chat rooms, and listservs. E-mail remains the single most popular application on the Internet. The degree to which a populist and democratic Internet survives and flourishes depends on larger social and political contexts. A revival of grassroots democracy in other arenas of American (or international) life—as happened in the 1960s—will reinforce grassroots democracy on the Internet (and not accidentally will make use of this medium to advance its causes).

The future remains uncertain. But it is clear that any history of the Internet will have to locate this story within its multiple social, political, and cultural contexts. This is particularly true since the Internet (in part because of its origins in the common language of binary digits and TCP/IP) seems to be emerging as a "meta-medium" that combines aspects of the telephone, post office, movie theater, television set, newspaper, shopping mall, street corner, and a great deal more.[75] Such a profound and complex development cannot be divorced from the idiosyncratic and personal visions of some scientists and bureaucrats whose sweat and dedication got the project up and running, from the social history of the field of computer science, from the Cold Warriors who provided massive government funding of computers and networking as tools for fighting nuclear and conventional war, and from the countercultural radicalism that sought to redirect technology toward a more decentralized and nonhierarchical vision of society.

The Road to Xanadu

Public and Private Pathways on the History Web

O N AUGUST 24, 1965, THEODOR NELSON PRESENTED a paper to the Association for Computing Machinery national conference in which he introduced the word "hypertext" to refer to "a body of written or pictorial material interconnected in such a complex way that it could not conveniently be presented or represented on paper." Nelson, who had started musing about this sort of associative thinking and linking as a Harvard University graduate student in 1960, viewed "hypertext" as an integral part of an imagined globally interconnected library and publishing system that would "grow indefinitely, gradually including more and more of the world's written knowledge" and "have every feature a novelist or absent-minded professor could want, holding everything he wanted in just the complicated way he wanted it held, and handling notes and manuscripts in as subtle and complex ways as he wanted them handled."[1]

Two years later, while working at the publisher Harcourt Brace, Nelson— an inveterate coiner of terms whose own Web page lists sixteen words or phrases that he claims to have introduced into general use—started to describe his global library as "Xanadu." "For forty years," Nelson wrote recently, "Project Xanadu has had as its purpose to build a deep-reach electronic literary system for worldwide use and a differently-organized general system of data management."[2]

Nelson's grand vision of a universal library and publishing system has come in for its share of derision. In 1995, the *Wired* magazine writer Gary Wolf devoted twenty thousand words to detailing what he called "The Curse of Xanadu." "Nelson's Xanadu project," he wrote, "was supposed to be the universal, democratic hypertext library. . . . Instead, it sucked Nelson and his intrepid band of true believers into what became the longest-running vaporware project in the history of computing—a 30-year saga. . . . [an] amazing epic tragedy. . . . [and] an actual symptom of madness." Nelson responded angrily to Wolf's profile, but he has also hinted that he views Xanadu as an impossible dream. He took the word from the imaginary home of Kubla Khan in Samuel Taylor Coleridge's uncompleted poem of the same name; Orson Welles (one of Nelson's heroes) used the same word for Citizen Kane's extravagant, uncompleted mansion.[3]

And yet, just five years after Wolf's obituary for Xanadu, the dream of a universal hypertextual library seems less like the narcotic imaginings of Samuel Taylor Coleridge or the fantasies of Ted Nelson than a description of a multibillion-dollar industry called the World Wide Web.[4] Even those of us whose professional calling requires us to think soberly about the distant past need now to consider whether such a contemporary development will reshape the ways we research, teach, and write history. Can professional historians look forward to a future in which they can access all the documentary evidence of the past with the click of a mouse? How far have we already come toward reaching that dream?

Not far enough yet. Even Nelson's 1965 paper on hypertext—quite relevant to anyone interested in the Web, which has hypertext as its most basic protocol—is not yet online. And any reader of this essay could come up with long lists of crucial historical sources only in physical libraries and archives. Still, a startling number of primary and secondary sources important to American historians have suddenly appeared online in the less-than-a-decade history of the World Wide Web. Indeed, so rapid has been the growth of the "History Web," as we will call that virtual world within a virtual world, that it cannot be readily surveyed within a single essay. I focus here instead on some of the *general trends* in the growth of the History Web over the past five years, especially its emergence as a rich online archive of primary and secondary sources, a Xanadu, in Nelson's words. What sources are now on-

line? What is the range and quality of this virtual archive? Even more important, *who* has put them there and *who* can use them?

Asking such questions inevitably leads us to wonder about the past, present, and future of one of the Internet's most celebrated qualities—its open
and *public* character. As the History Web has grown, it has also become more
complex. Many of the most important resources are now "hidden" from
view in databases not readily accessible by such Web search engines as Google
and AltaVista.[5] In addition, while many of the creators and owners of Web
content still come from what could be broadly called the public sector—
whether grassroots enthusiasts, grant-funded university-based projects, or
government agencies such as the Library of Congress—private corporations
(giant information conglomerates selling their wares to libraries, entertainment corporations trying to turn the Web into an advertiser-supported medium, and Internet startups with a range of business plans) are coming to
control some of the most valuable real estate on the History Web. Such private control raises questions about what history we will see on our computer
screens and who will be able to see it. If the road ahead leads to Xanadu.com
rather than Xanadu.edu, what will the future of the past look like?

One, Two, Many History Webs: Surface and Deep, Public and Private

Rapidity of change is a new technology cliché. "The Internet's pace of adoption," observes a United States Department of Commerce report, "eclipses all
other technologies that preceded it. Radio was in existence thirty-eight years
before fifty million people tuned in; TV took thirteen years to reach that
benchmark. . . . Once it was opened to the general public, the Internet
crossed that line in four years." In just the past five years, the percentage of
the United States population online has more than tripled from 14 to 44
percent. The "Web Characterization Project" of the OCLC (Online Computer Library Center, Inc.) reported 7.1 million unique Web sites in October
2000, a 50 percent increase over the previous year's total and almost a fivefold
increase since just 1997. Over that time, the Web has almost entirely displaced
other media—especially CD-ROMs—for presenting digital content. Conventional search engines such as Google currently index more than 1.3 billion
Web pages. Peter Lyman and Hal R. Varian estimate that in 2000 the World

Wide Web consisted of about 21 terabytes (a terabyte is 1,000 gigabytes) of static HTML (hypertext markup language) pages and was growing at a rate of 100 percent annually. But increasingly Web "pages" only come into existence as the result of specialized database searches, and those Web-based databases do not turn up in standard Web searches. BrightPlanet Corporation, whose Lexibot software indexes some of the searchable databases not readily accessible by conventional search engines, claims that this "invisible" or "deep" Web (in contrast to the "surface" Web found by the search engines) contains nearly 550 billion individual pages.[6]

How much has the History Web changed? No time machine can take us back to the Web of 1995 or 1996 and run comparative searches with today. One imperfect benchmark comes from searches that my colleague Michael O'Malley and I did in the fall of 1996 while writing an article on the History Web. Running the same searches in the same search engine (AltaVista) returns more than ten times as many "hits" today as four years ago—thereby greatly outpacing the overall growth of the Web and even "Moore's law," which predicts that computing power will double every eighteen months. We had 64 hits on William Graham Sumner, 300 on Eugene Debs, and 700 on Emma Goldman in 1996; the comparable figures for November 2000 were 716, 2,971, and 8,805.[7]

The quality of those "hits" improved as well. Four years ago, those looking for Debs on the Web might find some basic biographical information about the socialist leader, but the most interesting insights were how Debs fits into contemporary American life—how different groups (from the Democratic Socialists of America to the National Child Rights Alliance) and individuals (from local activists to Ralph Nader) made use of Debs's past in late-twentieth-century America. Now, however, the Web contains not only up-to-date biographical and historical treatments but also a gallery of images, state-by-state figures on Debs's presidential votes, guides to archival collections, and a substantial body of primary sources—at least a dozen different speeches or articles by Debs and another half dozen contemporary accounts of him.

Such raw Web searches do not, however, capture the fullness of the History Web since they do not generally measure the deep Web. For historians, the most notable of such databases are the more than 90 collections gathered under American Memory, the online resource compiled by the Library of Congress's National Digital Library Program (NDLP). Four years ago, American Memory had some staggering archival riches, but now the collection has grown at least fivefold and includes more than 5 million items— ranging from 1,305 pieces of African American sheet music to 2,100 early baseball cards. Visitors can examine 117,000 FSA/OWI (Farm Security Administration/Office of War Information) photographs, 422 early motion pictures and sound recordings of the Edison Companies, and 176,000 pages of George Washington's correspondence, letter books, and other papers. Library staff will soon place online another 30 collections, including such eagerly awaited resources as the thousands of ex-slave narratives of the Federal Writers' Project.[8]

Whereas four or five years ago history materials on the Web were most useful for teaching, the depth of such collections as American Memory means that historians can now do serious scholarly research in online collections. With more than 200,000 photographs now available in American Memory, anyone studying the history of American photography would need to visit the NDLP. Moreover, the digital format makes possible modes of research that are possible in other media but much more difficult. Take, for example, the old, but still much debated, question of George Washington's religious attitudes. Using the online version of the Washington papers, the historian Peter R. Henriques showed not only that Washington never referred to "Jesus" or "Christ" in his personal correspondence but also that his references to death were invariably "gloomy and pessimistic" with no evidence of "Christian images of judgment, redemption through the sacrifice of Christ, and eternal life for the faithful."[9]

Washington's dark thoughts on death are filed away in the deep Web of such databases as the vast American Memory collection not accessible by conventional Webwide search engines; Henriques's thoughts on Washington (published in print in *Virginia Magazine of History and Biography* but online through Bell & Howell's ProQuest Direct and EBSCO's World History Full TEXT), however, reside in a vast terrain that even BrightPlanet does not fully measure—what we will call the private Web. These are the growing number of online resources only available to paying customers. OCLC's data indicate that the growth of the public Web is slowing at the same time that private, restricted Web sites have gone from 12 to 20 percent of the total Web.[10] Whereas the surface and deep Webs, which together we will call the public Web, contain enormous numbers of primary documents, the private Web abounds in the secondary sources crucial to historical work.

For example, most historians know about JSTOR (Journal Storage: The Scholarly Journal Archive), which includes, in its 5-million-page database of

117 academic journals, the full text of 15 different history journals, most of them running from their inception up to 1995. Many of the nonhistory journals, for example, sociology, economic, and political science journals from the early part of the twentieth century, constitute primary sources of great interest to American historians. Searching JSTOR for Eugene Debs in history journals yields 81 articles, but expanding to other journals gives us another 61 articles, including such significant contemporary sources as John Spargo's "The Influence of Karl Marx on Contemporary Socialism" in the 1910 *American Journal of Sociology*. The word search capabilities of JSTOR also facilitate a kind of intellectual history that cannot be done as easily in print sources. Say you want to trace the changing reputation of Charles Beard in the historical profession; the 191 articles in JSTOR that mention Beard provide an invaluable starting point. Historians of language are already having a field day playing with such massive databases. The librarian and lexicographer Fred Shapiro, for example, has uncovered uses of such phrases as "double standard" (1912), "Native American" (for American Indian, 1931), and "solar energy" (1914) that predate citations in the *Oxford English Dictionary* by decades.[11]

JSTOR lacks the scholarship of the past five or six years, but online databases from Johns Hopkins University Press's "Project Muse" and the "History Cooperative" increasingly provide that as well. Although the History Cooperative, JSTOR, and Muse all restrict access to subscribers, they have emerged under nonprofit auspices. But increasingly important online collections of historical data are in the hands of commercial vendors such as Bell & Howell and the Thomson Corporation, which have vast archives of scholarly publications and primary sources, and Corbis, with its unparalleled archive of historical images. These are the exemplars of the private History Web—a growing realm both under corporate control and accessible only to paying customers.[12]

Everyone a Web Historian: Grassroots History Online

Despite the growing significance of the private History Web, the greatest energy over the past decade has actually been in the public Web—public in the sense of both its open access and its control by individuals, nonprofits, or government agencies. Indeed, an astonishing grassroots movement has fueled its enormous growth. Over the past five years, academic historians, history teachers, and history enthusiasts have created thousands of history Web sites. No one has managed a definitive count of these Web sites, although Yahoo!'s United States history directory includes more than 4,500 sites—a fivefold increase since 1996. My own Center for History and New Media maintains searchable databases of more "serious" history Web sites and has indexed more than 2,100 of them.[13]Although perhaps one-third of history Web sites have .com addresses (signifying the "commercial" domain in contrast to .edu, .org, or .gov), most of those are actually set up by individuals using free space (albeit festooned with banner and pop-up ads) provided by such companies as AOL (America OnLine), Geocities (a part of Yahoo!), CompuServe (an AOL subsidiary), Lycos, or Prodigy. To a surprising degree, then, history Web sites come from both academics and amateurs who have posted historical material online primarily as a labor of love—the original meaning of amateur.

Civil War enthusiasts, not surprisingly, have brought some of the same passion to presenting history online that they regularly display at Civil War reenactments. "Some days," observes *Choice*, the journal of academic libraries, "it appears that the Internet consists of equal parts Star Trek, stock market reports, soft-core pornography—and Civil War sites." And the historians William G. Thomas and Alice E. Carter have recently filled a two-hundred-page book surveying the Civil War on the Web, "a guide to the very best sites." Although many of these sites come from large institutions such as the Library of Congress, the National Park Service, and the Virginia Center for Digital History (with which Thomas and Carter have been associated), hundreds of passionate and dedicated amateurs have created remarkable sites without any outside financial or institutional support. Thomas R. Fasulo, an entomologist, has, for example, assembled an immense archive on the battle of Olustee (the largest Civil War battle in Florida)—more than forty official reports, fifty firsthand reminiscences in letters, articles, and books, and detailed coverage of all the units participating in the battle. The reenactor Scott McKay has developed an equally massive site on the Tenth Texas Infantry filled with rosters, casualty lists, ordnance records, battle reports, reminiscences, and personal letters.[14] To be sure, Civil War enthusiasts such as Fasulo and McKay flourished well before the emergence of the Web, but the Internet has made their passions visible and accessible to a much wider audience.

Genealogists have similarly found the Web a welcoming arena for engaging in their passion for the past. The USGenWeb Digital Library has mobilized hundreds of local volunteers to create online transcriptions of census records, marriage bonds, wills, and other public documents. The Family History Library of the Church of Jesus Christ of Latter-day Saints (the Mormon Church) has thrown open its massive genealogical databases, including the Ancestral and Pedigree Resource files (a database of family trees submitted to the Family History Library) and the International Genealogy Index (a name index of records collected by church members)—660 million names in all—the fruits of more than a century of Mormon genealogical work.[15]

Family historians have visited such sites in amazing numbers; the Mormon Church's site attracts 129,000 visitors per day, an annual rate of close to 50 million. Online resources have drawn tens of thousands more Americans into the already popular practice of tracing family roots—the most common form of historical research in the United States. Significantly, the Internet's greatest impact may lie in connecting people in common pursuit of their roots, allowing them to share information on common ancestors or to help out fellow genealogists by investigating a local lead. The Mormons alone sponsor 137,000 collaborative e-mail lists to facilitate research exchanges. While the Web has served largely as a publishing and archiving medium for already committed Civil War enthusiasts, it has brought new participants to genealogy by making the sources for family history more readily available. Print authors have even noticed the popularity of Web-based genealogical research; at least a dozen published guides—including *Genealogy Online for Dummies*—offer advice to enthusiasts.[16]

The breadth of this grassroots effort becomes clear when we look at who has posted a random selection of historical documents online. I pulled Diane Ravitch's anthology *The American Reader: Words That Moved a Nation* off my shelf and found online fifteen of the twenty documents (many of them far from mainstream) in her chapter "The Progressive Age." Teachers constituted the largest group of people who have made these documents publicly available—a communications professor at the University of Arkansas posting Elizabeth Cady Stanton's "Solitude of Self," a community college instructor in Ohio providing the Niagara Movement Declaration of Principles, a Hartsdale, New York, high school teacher digitizing M. Carey Thomas's "Higher Education for Women."[17] But many others had little or no academic connection. A black organizer includes W. E. B. Du Bois's "Talented Tenth" essay on his Web site (Mr. Kenyada's Neighborhood) because he believes that Du Bois's vision "of our potential capacity to solve problems internally" provides the basis for a new "community-based activism." A German purchasing agent puts Joe Hill's "The Preacher and the Slave" on his History in Song Web pages that preserve songs from an American studies course he took at Johannes Gutenberg University a quarter of a century ago. The General Board of Discipleship of the United Methodist Church publishes "Lift Every Voice and Sing," by James Weldon Johnson and J. Rosamond Johnson, with the suggestion that congregations "sing this hymn in worship on a Sunday in February [2000], and celebrate its one hundredth anniversary." The amateur poet Kevin Taylor's Web site includes Alice Duer Miller's pro-suffrage verse "Evolution" because "its message is as important and clear today as it has always been," and Miller "is also the author of *The White Cliffs*—one of my favorite books." The Web takes Carl Becker's vision of "everyman a historian" one step further—every person has become an archivist or a publisher of historical documents.[18]

Many of these grassroots efforts are quite modest, poorly designed Web sites proffering one or two favorite documents with little historical context. But others have grown into massive archives. In early 1995 the graduate student Jim Zwick began posting a few documents on anti-imperialism, the subject of his Syracuse University dissertation, on the Web. Like most historians, Zwick had assembled his own personal archive; he realized that the materials gathered for scholarly research could be made public through the World Wide Web. Five years ago, Zwick was one of the Web history pioneers; now his efforts have expanded well beyond anti-imperialism into such topics as political cartoons and world's fairs and expositions and thousands of historical documents personally digitized by Zwick. The volume of material and the number of users have multiplied more than fivefold. Although Zwick's Web site (now called BoondocksNet.com) remains a one-person operation, its increasing scale has forced him to take ads and sell books in order to support the growing hosting and software costs. Zwick has blazed a path that many future graduate students may (and I think should) follow. Why not take the least visible and most private part of the scholar's work—assembling a body of primary documents—and make it public?[19]

The most massive grassroots Web history effort linked to scholars is, of course, H-Net: Humanities & Social Sciences OnLine. Well known to historians for the more than 100 specialized discussion lists that it sponsors, H-Net also has a major Web presence, which includes searchable archives of the list discussions. HNet has not been heavily involved in posting historical documents, but its archives are now themselves a significant primary source for the thinking of professional historians, as well as an eclectic reference source to important books and teaching tools. Its most profound impact, however, has been on modes of scholarly communication; since its lists include 60,000 subscribers in 90 countries, it has become an essential way for historians to find out about conferences, grants, jobs, and teaching resources. To some degree, it has also accelerated the pace of scholarly discourse. In 1998, for example, subscribers to H-Amstdy, a part of H-Net, extensively debated Janice Radway's presidential address to the American Studies Association before it had been published in *American Quarterly*. Hundreds of volunteer list editors keep H-Net going, although the energy of Executive Director Mark Kornbluh, who has been very successful in obtaining government grants and university support, has also been vital to its maintenance and growth. As a result, H-Net remains a free scholarly resource that is also open to interested participants from outside the academy.[20]

The greatest strength of the grassroots History Web—its diversity and its links to nonprofessionals—is sometimes its greatest weakness. While academically trained historians such as Zwick and the H-Net community have joined in the bottom-up effort, its overall amateur and eclectic quality obviously poses problems for those committed to professional standards. William Thomas, for example, pronounces Civil War history on the Web "anemic" as well as "healthy." Few sites, he notes, "advance new ideas about the history of the period"; most ignore the scholarly trend toward social history and focus relentlessly on generals and battles. Still worse, "many web sites broadcast old prejudices, ancient theories, and long-disproved arguments about the Civil War," especially the view that the war was fought over tariffs rather than slavery. One site argues, "conditions in northern factories were as bad or worse than those for a majority of slaves" and rejects as "simplistic" the idea that "the Civil War was fought over slavery."[21]

Even amateur sites that stick to presenting primary sources rather than historical interpretations do not always meet professional standards. Reenactors digitizing battle reports or labor organizers posting Joe Hill songs generally do not fuss about proofreading and copyediting. Nor are nonprofessionals inclined to worry about definitive editions, editing, or careful contextualization. There are at least sixteen different online versions of Elizabeth Cady Stanton's well-known speech "Solitude of Self"; they provide conflicting dates on which she gave the speech and different bodies to whom she presented it. Paragraphing and punctuation vary widely, and some excerpt or even edit the speech without indicating the intervention. Only one provides a link to the Library of Congress, which has online a facsimile of a printed pamphlet version of the speech.[22]

Some documents found on the Web are, in fact, not "real" documents at all. At least three Web pages promise the "voice" of Eugene Debs, but the recording is actually that of Len Spencer, who recorded one of Debs's speeches around 1905.[23] More than two dozen different Web sites offer versions of what they call the "Willie Lynch speech of 1712," in which a British slave owner from the West Indies allegedly advises Virginia slave owners to control slaves through a strategy of divide and rule. Sometimes the sites add an introduction supposedly written by Frederick Douglass; others falsely describe Lynch as the source of the word "lynching." Despite the sites' repeated assurances about the speech's "authenticity," internal evidence readily betrays its twentieth-century origins. The language incorporates modern syntax, and the content focuses on current-day divisions such as skin color, age, and gender rather than ethnic and national divisions much more important in the early eighteenth century.[24]

To be sure, a careful search of the Web also turns up evidence of the dubious origins of the Lynch speech. Still, those sites that take the speech entirely at face value overwhelm the Web sources that dispute it. Anyone who simply searched for "Willie Lynch" on the Web would be more than ten times as likely to find evidence of the speech's "authenticity" than information that casts doubt. But the Web is unique in the way it offers entry into the world of information and misinformation in which most people operate and allows us to consider the significance and spread of such urban legends as the Willie Lynch speech, which are orally transmitted at such events as the 1995 Million Man March or the 2001 inaugural protests. The Web itself cannot be blamed for misinformation or misrepresentation; the Lynch speech, in fact, appeared in print as early as 1970. The Web increases our access to documents and information, both spurious and authentic. For both better and worse, the virtual archive of the Web distinguishes itself from traditional libraries and archives by its indiscriminating inclusion of the best—and worst—that has been known and said.[25]

Despite the abundant misinformation available online, the Internet is—somewhat paradoxically—a superb source for basic factual research, especially when used by those who are careful to determine source quality. My own rendering of the Willie Lynch story comes entirely from research in online sources. Although I have a substantial reference library at home, I now do most of my historical "fact checking" on the Web. I can find correct spellings, birth dates, battle deaths, and election results in online sources more quickly and more accurately than in most standard reference works. The key caveat, of course, is "careful to determine source quality," but most professional historians—and probably most advanced history students or most sophisticated general readers—possess this skill.

Deepening the Public History Web: Universities, Foundations, and the Government

While the largest number of Web sites with historical documents and content have emerged out of this eclectic, grassroots effort, the largest *volume* of historical documentation exists within the deep Web of online databases and the private Web of materials open only to those who pay. Both efforts share some basic similarities—massive scale and use of databases to organize the materials. But only paying customers can visit the private Web.

Surprisingly, enormous amounts of free online historical material have appeared in the past five years, and much more will appear in the next decade. These sites have primarily benefited from government or foundation funding or, in many cases, both. The most important project, the Library of Congress's National Digital Library, has spent about $60 million to put more than 5 million historical items online between 1995 and 2000—with three-quarters of the funding coming from private donations. Ameritech, the former Bell telephone company for the Midwest (now owned by SBC Communications), worked with the Library of Congress to provide $2 million for more than 20 digitization projects at libraries across the country.[26] The heavy corporate funding naturally raises the specter of probusiness bias in what gets digitized. The AT&T Foundation, for example, has supported the digitizing of the Alexander Graham Bell Family Papers. The Reuters America Foundation was probably more likely to support the digitizing of the George Washington Papers than the records of the National Child Labor Committee. Nevertheless, Ameritech has, for example, funded the Chicago Historical Society's efforts to bring its collection of Haymarket affair materials to the Web.

The National Endowment for the Humanities (NEH) has also supported many important projects, particularly favoring those with an educational mission and focus on particular topics. The well-known Valley of the Shadow Project at the University of Virginia brings together a stunning archive of documents about two nearby counties (Augusta County, Virginia, and Franklin County, Pennsylvania) on opposite sides during the Civil War era. Already a major Web destination in 1996, its collection of letters, diaries, newspapers, censuses, and photographs has multiplied tenfold in just the past four years. The Valley of the Shadow is remarkable not just for its depth and sophistication but also because it has no physical counterpart. Edward L. Ayers, William G. Thomas, and their collaborators have literally created an archive that did not previously exist by hunting down and digitizing documents found in both public repositories and private hands.[27]

The New Deal Network (NDN), another NEH-funded project, has similarly created a new, virtual archive, with more than 20,000 photographs, political cartoons, and texts (speeches, letters, and other documents) gathered from multiple sources. Sponsored by the Franklin and Eleanor Roosevelt Institute and led by Tom Thurston, the New Deal Network lacks the comprehensiveness of the Valley of the Shadow, but it offers a remarkable resource for anyone teaching about the 1930s and 1940s. History Matters: The U.S. Survey Course on the Web, the product of my own Center for History and New Media and the American Social History Project and funded by NEH and the Kellogg Foundation, has digitized hundreds of first-person historical documents and contextualized them for use in high school and college classrooms.[28]

In contrast to the "invented archives" represented by the Valley, NDN, and History Matters, Documenting the American South opens up an existing archive—the University of North Carolina at Chapel Hill's unparalleled Southern collections—to remote students and scholars. Funded by various grants (from NEH, Ameritech, and the Institute of Museum and Library Services), Documenting the American South organizes thousands of documents (largely texts) around such specific topics as "Southern Literature," "First-Person Narratives," "Slave Narratives," "The Southern Homefront, 1861–1865," and "The Church in the Southern Black Community."[29]

The National Science Foundation (NSF), with a budget thirty times that of NEH, has emerged as an important funder for "digital libraries" as a result of its interest in computing issues rather than in the quality of the content being provided. Whatever the motives, NSF has financed some projects of enormous interest to historians. Michigan State University's National Gallery of the Spoken Word (NGSW) is developing techniques for automatically searching large volumes of spoken materials, including, for example, thousands of hours of nightly TV news broadcasts. Historians may not care about the underlying computer science, but if the NGSW succeeds in creating a "fully searchable digitized database of historical voice recordings that span the 20th century," they will make extensive use of it in their teaching and research.[30]

Whereas NEH funding has largely supported the creation of digital projects for use in the classroom and NSF has concentrated on the intersection of computing and humanities problems, the Mellon Foundation has focused on library-related issues, especially preservation and storage. It has provided substantial funding to the Cornell and University of Michigan libraries to preserve and then make available a major library of printed materials published between 1850 and 1877 under the rubric of the "Making of America" (MOA). The University of Michigan portion of the collection alone will soon encompass more than 9,600 monographs, 50,000 journal articles, and 3 million pages—a significant portion of the library's imprints from those years.[31]

Like scholars using NDLP, those using MOA can find information previously available in theory but not necessarily in practice. Steven M. Gelber, who was researching the origins of hobbies, notes that he turned up "a treasure trove of data in a matter of a couple of days" that would have taken months to find through traditional research. He calls MOA "the most exciting thing I have seen in research since I first discovered Xerox machines in 1967 and realized I did not have to take notes anymore." This "is what I assumed the future of libraries would be but to be quite honest, I never believed I would live to see so much of the past put online in such an accessible form."[32]

Despite the enormous value of the MOA and similar projects, some cautions are in order. Some object that such efforts are a form of burning down the village to save it, since most of the books will ultimately be discarded— both because they are cut up to be scanned and because the storage space is valuable. The novelist Nicholson Baker, for example, has sharply criticized earlier newspaper microfilming projects that have led to the similar destruction of paper copies of the newspapers. As the result of Library of Congress microfilming efforts, for example, libraries across the country dumped their hard copies in the belief that there was now a standard, comprehensive microfilmed version of newspapers that could be reproduced, ordered, and consulted. But Baker argues that the anomalies and holes (missing issues, pages, etc.) in the Library of Congress collection have now become permanent holes in some newspaper records because of the ensuing destruction of holdings in other libraries.[33] Baker and others also note the value of marginalia and other markings that get lost with the disappearance of paper copies as well as the difficulties of fully reproducing images such as nineteenth-century engravings in digital form. Librarians, on the other hand, argue that books and newspapers printed on acidic paper were crumbling and that microfilming or digitizing offers the only practical alternative and the only way to supply "the most content to the most people in a cost-effective manner." While some scholars will bemoan the loss of tangible, historical evidence in the transition from paper to digital images (just as they mourn the disappearance of the card catalog), many others will benefit from their ability much more readily to access the volumes in the MOA collection, many of which are not in a standard university library, and even more the possibility of searching them by words in the text rather than just by title.[34]

Indeed, the incredible ease of using these newly digitized works may actually pose a problem for future historical work. The MOA collection largely draws from books from Michigan's remote storage that had rarely been borrowed in more than 30 years. Yet the same "obscure" books are now searched more than 500,000 times a month. Will digitization create a new historical research canon in which historians resort much more regularly to works that can be found and searched easily online rather than sought out in more remote repositories? Years ago, the *New York Times* ran an advertisement with the tagline "If it is not in the *New York Times* Index, maybe it didn't happen." Could we arrive at a future in which, if it is not on the Web, maybe it didn't happen?

Such concerns aside, these grassroots, government, and nonprofit efforts have begun to deliver, as Gelber observes, "what people have been talking about for ten years—a genuine electronic library, or at least an electronic archive." Historians will spend years examining these digital sources and will not readily exhaust their possibilities. Although the Founding Fathers may be better covered in these resources than labor or feminist militants are, the Web in fact now offers material stretching across the broad range of topics that interest contemporary historians. The always precarious state of the public sphere in contemporary America poses one crucial peril for the continued expansion of this burgeoning free archive. For example, the budget of NEH, the most important funder of humanities work, has declined (in real terms) by about two-thirds in the past twenty years.[35] And in the past several years, it has had to fight for its survival. NEH may now face further threats with a Republican president and Congress who traditionally have not been sympathetic to the public sector.

Despite the great success of American Memory, which receives eighteen million page views per month and has brought primary sources into K–12 classrooms across the country, the Library of Congress seems to be shifting away from its focus on putting its historical collections online. A report by the National Research Council in the summer of 2000 criticized the library for, in effect, paying too much attention to historical sources and not enough to recently created "born digital" materials such as Web sites and electronic journals and books. James O'Donnell, vice provost for Information Systems and Computing at the University of Pennsylvania who chaired the committee producing the report, told the *New York Times*: "Digitizing your analog material is less urgent. . . . If you don't do it this year, it'll still be there in five years, and you could do it then. Digital information that you're losing is probably lost forever."[36] If the Library of Congress turns away from the massive digitizing efforts of the past five years, American Memory may turn out to be a forgotten memory from the late twentieth century.

Moreover, most of the government or foundation funding has been significantly enhanced by university support (another part of the endangered public sector) and by substantial infusions of sweat equity from digital pioneers. When the creation of online archives becomes routine, will that university and volunteer support remain available? In other words, is there a stable basis for the continued funding of public sector efforts to create a public, free historical archive?

The continuing erosion of the "public domain" further threatens the public Web. Copyrighted material previously entered this intangible realm of unrestricted use after a twenty-eight-year term renewable once, or a maximum of fifty-six years. In 1976, the copyright law narrowed the public domain by lengthening most existing copyrights to seventy-five years. As a result, the only large bodies of materials for the years after 1923 (the year after which copyright covers most work) are government documents such as the WPA (Works Progress Administration) life histories or the FSA photographs. The Sonny Bono Copyright Term Extension Act of 1998, which extended copyrights for an additional twenty years (in part due to the aggressive lobbying of the Disney Corporation, whose Mickey Mouse was scurrying toward the public domain) means that the copyright line will remain frozen at 1923 until 2018. Thus, Web surfers can easily read F. Scott Fitzgerald's *Tales of the Jazz*

Age (1922) but not *The Great Gatsby* (1925), which will not find its way online until 2020. The 1998 copyright extension delivered the single greatest blow to the creation of a free, public historical archive; yet historians were barely at the table when that act passed, crowded out by the high-priced suits from the big media conglomerates. Copyright restrictions are one reason for the persistence of fading digital formats such as CD-ROM. The two United States history CD-ROMs on which I have worked contain copyrighted materials that we could purchase permission to use in the CD-ROM but not on the Web.[37]

Selling the Past Online: Information Conglomerates and Internet Startups on the Private History Web

For historians, copyright protection has redlined not only much twentieth-century history but also most secondary literature out of the public Web. But because the problem involves rights and money, one solution similarly involves rights and money: companies that provide copyright digital content, charge for it, and then compensate rights holders out of their revenues. That said, the particular models for selling digital content vary widely as the corporations in the emerging "information business" scramble to evolve the most profitable business model.

The most common approach involves high-priced library-based subscriptions to digital content. Individual library subscriptions, which allow the library to provide the materials to all its patrons, generally cost thousands of dollars. The Virtual Library of Virginia (VIVA), which purchases electronic databases for the state's 39 public college and university libraries (a consortium arrangement increasingly common in this environment), currently spends more than $4 million per year for electronic subscriptions, and individual libraries in the consortium are spending thousands, if not millions, more.[38] Annual subscriptions to periodical databases such as ProQuest Direct and Expanded Academic ASAP (EAA) typically run around $30,000 to $50,000 for colleges and universities.

Other vendors sell digital content on an item-by-item basis—"by the drink"—instead of by subscription. Northern Light, which modestly aspires (in the words of its chief executive officer) "to index and classify all human knowledge to a unified consistent standard and make it available to everyone in the world in a single integrated search," offers more than 700 full-text publications (including a number of history journals) on a per-article basis. You can, for example, get Howard Zinn's article in the *Progressive* on "Eugene V. Debs and the Idea of Socialism" delivered instantly to your Web browser for $2.95. Contentville, which has more of the feel of a magazine (it was founded by Steven Brill, who made his millions with such publications as *American Lawyer*), offers a smaller selection of articles at similar prices as well as primary source documents such as speeches and legal documents. Prominent academic experts such as Sean Wilentz and Karal Ann Marling recommend the best books on "American Politics since 1787" and "Popular Culture," and contributing editors share their favorite Web sites.[39]

The vast image library controlled by Corbis, the company owned by the Microsoft founder Bill Gates, offers up the most massive historical database available on the pay-per-drink basis. Corbis has swallowed up many of the world's largest image collections, including the Bettmann Archive and the French photo firm Sygma, and has licensing arrangements with leading photographers and repositories around the globe (from the National Gallery in London to the State Hermitage Museum in St. Petersburg). It also represents another example of the trend toward massive concentration in the digital environment. Increasingly, the world's images are coming under the control of just two giant Seattle-based firms—Corbis and Getty Images, owned by the oil heir Mark Getty. Both aspire to be, as a Corbis ad says, "your single source for an array of diverse images"—"The Place for Pictures Online," in its trademarked phrase. More than 2 million of Corbis's 65 million images are digitized and available through a fast search engine. Anyone who has done photo research for a book or article will appreciate the ability to sit at home and browse through this incredible collection—17 superb photos of Eugene Debs, for example. You can look for free, but using the images (emblazoned with "corbis.com" in the online version and protected with digital watermarks) comes with a price tag that escalates as you move up from a digital image for your personal Web page ($3) to a glossy print for your wall (starting at $16.95) to an image that you can publish in a book (generally $100 or more).[40]

Corbis's charges reflect copyrighted images in many cases, but in others they rest on the company's ownership of an image published widely in the pre-copyright era and available for free if you can get a copy from a less fee-hungry source such as the Library of Congress. You can pay Corbis $3.00 for a digital image of Walker Evans's photo of the "Interior of a Depression-Era

Cabin" or download a higher quality version of the same image in American Memory for free. American Memory also provides a fuller identification and contextualization of the photo, since its goals are educational and scholarly rather than just pecuniary. Similarly, you can purchase Eugene Debs's 1918 Canton, Ohio, speech, which helped land him in prison for sedition, from Contentville for $1.95 or you can pick it up for free on at least four different Web sites.

Costs aside, these online databases are already revolutionizing the way historians do their research. Most familiar to historians are the massive bibliographic databases such as America: History and Life and the Arts and Humanities Citation Index. Once upon a time (that is, five or six years ago), historians searched through annual bound volumes to develop bibliographies. Now they typically do these searches quickly and at their own convenience. After assembling a bibliography, historians used to search for and copy articles. But now they can find the full text of a surprisingly wide selection of secondary works online.

The major online sources for full-text journals—Bell & Howell's ProQuest Direct, the Thomson Corporation's Expanded Academic ASAP (EAA), and EBSCO—offer thousands of journals, including dozens of major historical journals, generally from 1989 to the present.[41] Despite some gaps such as most state historical society publications, these databases contain a large percentage of the journal literature of the 1990s that historians would need to consult. Two other nonprofit, but still gated, resources—Project Muse and the History Cooperative—fill in some important gaps in what ProQuest and EAA offer. For still older sources, JSTOR (also available only through hefty library installation charges as well as an annual maintenance fee) provides comprehensive coverage, albeit for a smaller set of journals.

As yet, historical monographs cannot be found in cyberspace as readily as journals can. But perhaps not for long. Questia Media, Inc., backed by $130 million in venture capital, has created an online liberal arts library of 50,000 scholarly books, which they hope will increase to a quarter million volumes by 2003—what they call the "world's largest digitization project." Taking an approach different from that of ProQuest and EAA, Questia intends to sell subscriptions for $19.95 per month to "time-crunched" students, who they believe (in the face of some reasonable skepticism) will pay for access to materials that will help them write their papers more quickly. At least in history classes, the investment may not pay off: although Questia has more than

9,000 history titles, not a single one of the 10 history monographs that United States historians, in a *Journal of American History* survey, listed as "most admired" can be found on the online library's shelves. Its competitors, NetLibrary (with more than $100 million in venture capital and 25,000 books already online) and Ebrary.com, have still other business models. NetLibrary sells libraries electronic copies of books that can only be accessed by one person at a time; if someone has "checked out" the book, then no one else can "take it out." It markets its 25,000 books in different groupings ranging from the 618-title "business school collection" at an average price of $40 per volume to 126 volumes on "Countries, Cultures, and Peoples of the World" to 214 volumes of "Cliffs Notes" (the actual literary works are generally thrown in free since they are in NetLibrary's collection of 4,000 public domain books). Ebrary, by contrast, allows users to browse books without charge but requires payment for printing or copying a portion of a book.[42]

Not all pay services offer copyrighted content. Some serve public domain content but charge in an effort to recoup their digitizing costs. One of the pioneers in this has been HarpWeek, a personal project of John Adler, a retired businessman with an interest in nineteenth-century American history. While most digitizing projects rely on "keyword" searching of the full text, Adler has employed dozens of indexers to read every word in *Harper's Weekly* and examine every illustration and cartoon to create a human index of the full run of the magazine from 1857 to 1912. That labor-intensive indexing means, for example, that HarpWeek offers better image searching than many other online sources since the brute power of keyword searching brings much greater rewards in historical texts than in images. Adler has created an extraordinary research resource for nineteenth-century historians, although an expensive one—the first twenty years, now available, retail for close to $35,000.[43]

We can glimpse the outlines of a still more remarkable project—the full text of the *New York Times* for the years 1851 to 1923. The "Universal Library" at Carnegie Mellon University (with aspirations similar to Nelson's Xanadu project and support from Seagate Technology) is scanning the entire public domain era of the *Times*, which it will make available for free online reading. At the same time, it is using optical character recognition to turn the *Times* into searchable text, although the quality of the result remains uncertain at the moment. The Universal Library plans to offer free views of the page images but to charge for access to the searchable text—perhaps $40 for lifetime subscriptions. At the moment, the vision is more exciting than the implementation—you can't search yet, and the scanned microfilm provided for 1860–1866 includes a number of unreadable pages.[44]

The plan of the university-based Universal Library to charge subscriptions suggests a type of History Web site that sits uneasily between the "public" and "private" categories that we have been using. Like JSTOR and Project Muse— both of them nonprofit ventures that have received substantial support from the Mellon Foundation—it is "public," rather than private, in its ownership, control, and eschewing of profit. Yet, it is (or will be) "private" in its restriction of full access to those who pay. Despite their foundation funding, groups such as JSTOR and Project Muse argue—quite reasonably—that they need income to sustain their operation, to add new journal articles, and to maintain the service. Thus, they charge substantial subscription fees to libraries. Unfortunately, when nonprofits enter the private Web, they not only restrict access but also incur substantial costs; JSTOR and Project Muse spend a considerable part of their income not to create or post content, but to market their services and keep out unauthorized users. Michael Jensen, who helped develop Muse, estimates that "over half of the costs of the online journals project was attributable to systems for *preventing access* to the articles."[45]

Moreover, even where publication, preservation, or distribution is turned over to a nonprofit such as JSTOR or Project Muse, scholarly authors and journals are still giving up control over presentation and access to a separate entity. The History Cooperative—a partnership of the University of Illinois Press, National Academy Press, the Organization of American Historians, and the American Historical Association—has pioneered the alternative idea of a "cooperative" in which scholars and scholarly organizations will retain a say over these questions.[46] Historians from these professional societies and their journals felt that this arrangement would allow them, for example, to offer to make their electronic journals as widely available as possible. Hence, while the electronic *Journal of American History* and *American Historical Review* will only be available to subscribers, there is no additional subscription charge to individuals or libraries for access. Having a say in a cooperative also makes it easier to experiment with one of the key questions facing scholars— will digital environments allow us to present our scholarship in new—and better—ways?[47] In the end, the measure of success of scholarly and nonprofit societies is how they improve scholarship and society, not how much revenue they generate.

Some argue that, given these larger social and scholarly goals, scholars should move toward total, free access to the fruits of scholarship, which is, after all, mostly publicly funded in the first place. In 1991 Paul H. Ginsparg, a physicist at the Los Alamos National Laboratory, created arXiv.org e-Print archive, which has become an open repository of more than 150,000 "pre-prints" (nonpeer reviewed research papers) in physics, math, and related fields. "E-print" archives in psychology, linguistics, neuroscience, and computer science similarly offer electronic preprints on a free access basis. The Open Archives Initiative advocates expanding these efforts so that they will be "interoperable" (for example, allowing easy searching across multiple archives); include peer-reviewed work; and ultimately form the basis of a "transformed scholarly communication model." The computer scientist Stevan Harnad, one of the most aggressive promoters of such open systems, envisions a future in which "the entire refereed literature will be available to every researcher everywhere at any time for free, and forever."[48] Thus far, scientists have dominated such open scholarly archive experiments. It remains a question whether they are easily transferable to the humanities, which lack the same preprint traditions and where speed of publication is much less important. Moreover, the extraordinarily high prices of commercially published science journals have further driven these efforts. No one worries about putting commercial science publishers out of business. But the losers in the demise of the scholarly history journals will be university presses and scholarly societies.

If scholarly societies such as the Organization of American Historians are to survive in a world where all scholarly information is free, they will need to come up with alternative revenue models to support their operations. One promising approach to resolving the contradiction between free public access and continued revenue to support scholarly editing and publication has been pioneered by the Open Book project at the National Academy Press (NAP), which has been led by Michael Jensen, who has also been a key figure in Project Muse and the History Cooperative. NAP, the publishing arm of the National Academy of Sciences, has put its entire front list and much of its back-list online for free in a page image format. Ironically, giving this material away has actually increased NAP's sales because people now order books that they have browsed online but want to own in a hard copy. Moreover, the book itself—indexed by Web search engines—becomes its best advertisement. Jensen, thus, argues that "free browsing, easy access, and researcher-friendly publication first, and sale second" is "much more in keeping with the role of a noncommercial publisher" and its mission of doing "the most good for society as possible within the constraints of our money."[49]

Who Owns the Past Online? Access and Control on the Private History Web

These massive projects, whether public or private, will surely transform historical research and ultimately writing. Those who received their Ph.D.s before 1990 will probably spend the rest of their careers regaling graduate students with tales of how "in my day, we spent hours turning microfilm readers looking for relevant newspaper articles." Given the enormous gift that commercial digitization is bestowing on the historical profession, it seems a bit churlish to look this particular gift horse in the mouth

Churlish, but surely necessary. Once we get over our excitement about the digital riches on our screens or the new modes of research being opened up, we need to think about the price tag. To be sure, in most of the emerging models, libraries rather than individual researchers are paying that fee. Still, that money is not appearing magically; it is draining other parts of library budgets. One part of the budget that is being sucked dry is that for purchasing real, not virtual, library books, especially scholarly books. To be sure, the main villains in the current crisis in scholarly publishing are the commercial vendors who charge rapacious prices for science, technology, and medicine journals. Libraries that pay $16,344 annually to subscribe to Reed Elsevier's *Brain Research* cannot afford as many history monographs as they once purchased, a fact that both scholars and university presses are painfully confronting. But electronic resources are also squeezing library budgets—they now consume 10 percent of library materials budgets, compared to only 25 percent for monographs.[50]

The digital library fees also generally flow into the hands of publishers and especially commercial aggregators rather than authors. Freelance writers have sued newspapers and magazines for including their work without permission (or compensation) in databases marketed by Lexis-Nexis (Reed Elsevier) and Bell & Howell. And book publishers have been slow to decide what portion of e-book revenues they are going to share with authors.[51]

In addition, the appearance of these gated databases poses a particular problem for independent scholars not affiliated with academic institutions. If they happen to live near a major public library, they can often access the databases within the walls of that library. But they do not have the convenience available to most university-based historians of using these resources from their own homes.[52]The same problem faces those affiliated with smaller institutions that cannot afford the hefty subscription fees. Some scholars, however, now have enhanced access to resources; in Virginia, VIVA's statewide subscriptions give historians at community colleges and underfunded traditionally black colleges access to the same electronic resources as faculty at the well-endowed University of Virginia. Nevertheless, signs of an academic digital divide loom not only between institutions but also within them. For example, law school students and faculty generally have access to the complete Lexis-Nexis database (with considerable resources for historians), which is generally closed to other parts of the university. Of course, scholars affiliated with more affluent institutions (and parts of institutions) have always had advantages over their colleagues, and independent scholars have always faced barriers to access.

A more worrisome prospect has to do with the emerging economic structure of the information industry. Previously, publishing was a relatively decentralized and small-scale business with many different publishers, large and small. But online information providers, like many other "new economy" businesses, benefit from a powerful combination of economies of scale and "network effects." In the information business, the fixed costs (for example, software development) are the most important costs; once they are covered, it is not much more expensive to sell to 3,000 libraries than to 30. And "network effects"—the benefits of using a system increase as more people use it since, among other things, they will be familiar with its interface—mean that the biggest players will tend to get bigger. Whereas the factory-based economy favored oligopolies, the information economy is more likely to result in monopolies.[53]

Not surprisingly, then, the online vending of electronic data has already become concentrated into a very small number of hands. Four gigantic corporations—Reed Elsevier, EBSCO, Bell & Howell, and Thomson—are especially prominent in the provision of electronic content to libraries. Reed Elsevier, which focuses particularly on science journals, is less significant for historians (although it does sell Lexis-Nexis, the online data service vital to anyone writing on the recent past). The privately held EBSCO, which has $1.4 billion in annual sales, produces nearly 60 proprietary reference databases and full-text versions of more than 2,000 publications. Bell & Howell is a billion-dollar corporation, which acquired UMI (formerly University Microfilms International) in 1985 and Chadwyck-Healey (a leading provider of humanities and social science reference and research publications) in 1999. Its databases include over 20,000 periodical titles, 7,000 newspaper titles, 1.5

million dissertations, 390,000 out-of-print books, 550 research collections, and over 15 million proprietary abstracts. These resources constitute an archive that includes more than 5.5 billion pages of information—all of which is being converted into digital form (though not necessarily searchable text) under the "Digital Vault Initiative," which the company says will create "the world's largest digital archival collection of printed works." ("World's largest" is a popular claim in cyberspace.) Ultimately, Bell & Howell will offer online the full runs of at least 50 periodicals such as the *New York Times*, *Time*, and the *Wall Street Journal*. (Astonishingly enough, given the scale of the effort involved, Bell & Howell intends to create its own searchable edition of the *New York Times*, and its version will come up to the present rather than stop in 1923.)[54] The microfilm era in research, which Bell & Howell's UMI launched in 1938, will soon come to an end.

Bell & Howell's even larger rival is the Canadian Thomson Corporation, a "global e-information and solutions company" with close to $6 billion in annual revenues. Thomson's Gale Group sells thousands of full-text publications (including history journals) to libraries under the "InfoTrac" brand, which includes EAA. It also has extensive reference holdings, including works that historians regularly use (for example, from Macmillan Reference USA and Charles Scribner's Sons). More recently, it has bundled its various products as well as some licensed from other vendors into what it calls its "History Resource Center," billed as "the most comprehensive collection of historical information ever gathered into one source." Designed primarily for undergraduates and to be purchased by college or university libraries, it includes primary documents (from an archive accumulated by Primary Source Media, another Thomson subsidiary), encyclopedia articles, full-text periodicals and journals, maps, photographs and illustrations, overview summaries, a timeline, a bibliography, and annotated links to online special collections. These resources do not come cheap. Prices vary considerably depending on particular arrangements, but an annual license for two simultaneous users can run close to $12,000.

Bell & Howell and Thomson are involved in a dense web of connections with other online ventures. Thomson, for example, holds the largest stake in WebCT.com, which provides widely used software for placing courses online but bills itself more broadly as an "e-learning hub." WebCT has developed discipline-specific online communities with forums and other resources, including one in history. Part of the reason for Thomson's "strategic investment" is presumably to encourage the selling of custom course materials created by Thomson to students in courses managed through WebCT. Bell & Howell is also eyeing the lucrative textbook (or "courseware") market and has recently launched XanEdu, which repackages the materials that it sells to college libraries as ProQuest and sells them to students as electronic course packs and a subscription-based ($49.90 per year) "elibrary for college students, with targeted content and course-driven pre-selected searches" in such fields as history. For the K-12 and public library markets, Bell & Howell further repackages some of the same resources through BigChalk.com.[55]Bell & Howell and Thomson, thus, aspire to dominate not only university-based library reference publishing but also textbook publishing and education at all levels. In the new electronic environment, such previously separate enterprises potentially merge together into information "portals" or what XanEdu calls "the ultimate learning destination." Like Ted Nelson, from whom they may have borrowed their new corporate moniker, the folks at Bell & Howell dream big, promising that XanEdu will be a "utopia for the mind."

Advertising offers another road to a corporate-owned past. Some believe that the Web will emerge as the primary advertising venue of the future, replacing television and glossy magazines. In that scenario, "free" information would be served up in the same fashion as television offers "free" entertainment. Entrepreneurs and large corporations have launched dozens of Web sites aimed at making money off the provision of historical or educational information and services through advertising or marketing. Some, such as the HistoryChannel.com or Discovery.com, are spin-offs of existing print or cable operations. For example, The HistoryNet.com (billed as "where history lives on the Web") is the online companion to fourteen popular history (mostly military history) magazines, including *Civil War Times*, *Wild West*, and *Aviation History*. In addition to back articles from the magazines, it offers a daily quiz, "This Day in History," recommended Web sites (limited in coverage), online forums (not very active in the fall of 2000), and lists of history-related events and exhibitions—all accompanied by flashing banner ads.

Still other history-related sites are startups created directly for the Web. About.com (formerly the Mining Company), for instance, dubs itself the "Human Internet" and provides human "guides" to more than 700 different subjects, including "Women's History," "Twentieth-Century History," and 10 additional historical subjects. The guides, who generally have an undergraduate history degree, usually offer brief annotated links to Web-based materials, short essays of their own (often with some connection to current events), and online forums. The forums—most of them not especially active—include a homework help feature to which students post queries. (Judging from the answers, I doubt everyone will get an A.)

Many other Web startups have shared About.com's interest in tapping the education "market"—an expansive realm including teachers and students at multiple levels. During the Internet stock fever that raged through most of 1999 and early 2000, education dot-coms sprouted overnight as dreams of IPO (initial public offering) millions danced in the heads of entrepreneurs and venture capitalists. Typical were eCollege, a distance education company that raised $55 million in an initial public offering in December 1999, and Lightspan, a provider of "curriculum-based educational software and Internet products," including, it promises, lesson plans and source documents in history and other fields.[56] Lightspan went public at $11.625 per share in mid-February of 2000, and the stock more than doubled less than a month later.

So far the reality of the sponsored history and education sites has not matched the glittering promises, whether of immense profits or of illuminating content. Generally speaking, the nonprofit sites offer considerably better content. For example, 774 popular history articles available at The HistoryNet.com pale beside the thousands of scholarly articles offered at JSTOR. The richest materials at About.com are those from such sites as American Memory and the New Deal Network, which are presented framed beneath About.com's banner ads. H-Net and History Matters provide considerably more active discussion forums than does The HistoryNet or About.com. The History Channel's list of best history Web sites lists the site of the Eighteenth Louisiana Infantry Regiment but not Valley of the Shadow or the Library of Congress's collection of Civil War photographs—presumably because you must sign a partnership agreement with the History Channel and post its banner ad to get listed. One must view skeptically The HistoryNet's claims that it is "the Internet's largest and most content-rich history site" or About.com's boast that "our Guides know their subjects as well as anyone."[57]

Stock prices have been even more inflated than content claims, as the spring 2000 NASDAQ (National Association of Securities Dealers Automated Quotations) crash brutally revealed. About.com lost almost three-quarters of its stock value between March and April 2000; eCollege stock plunged 85 percent, and Lightspan plummeted to just above one dollar a share. "There are a lot more companies in the e-learning space than the education industry needs," acknowledged eCollege's chief executive officer, Oakleigh Thorne. Companies with real rather than virtual sources of revenue also began to wonder whether there really was a pot of gold at the end of the Internet rainbow. In November 2000, the privately held Discovery Communications dropped plans to spin off its Web unit and also dropped most of its Web workers—laying off 40 percent of the regular staff and 150 contract workers. "We cannot achieve near-term profitability from the Internet as a stand-alone business," explained the company president, Michela English. Part of the problem was that none of these sites was ever profitable; they simply lived off venture capital, IPO money, or the largess of wealthy corporate parents. Equally problematic was the drop in Internet advertising rates that accompanied the dive in Internet stocks and the realization by advertisers that few Web surfers (about 0.4 percent) were clicking on banner ads.[58] The fall in rates was part of a vicious cycle in which dropping stock prices soured advertisers on the Internet and then caused problems for startups, which—in a kind of Ponzi scheme—had artificially raised rates in the first place with their own advertising.

The collapse of dot-com stock prices and Internet advertising rates suggests that the future of commercially sponsored history on the Web may not be as rosy as some once believed. The history business has had its share of successes in the "real" world—from *American Heritage* magazine to the History Channel, from the History Book Club to heritage tourism—but it has never been a major American industry.[59] The past remains a realm in which nonprofits, volunteers, and enthusiasts dominate.

Still, as Susan Smulyan reminds us in her history of the commercialization of American broadcasting, broadcasters and advertisers, as well as listeners, viewed the viability of radio advertising with considerable skepticism. Someday Web advertising may be as "natural" and profitable as television commercials. The drop in Internet advertising *rates*, moreover, has not halted the continuing rise in the overall volume of Internet advertising.[60]And the bursting of the dot-com stock bubble has not slowed the growth in Internet use or even the increasing importance of the Web as a commercial venue. Whether or not history will turn out to go better with Coke (ads), the selling of digital information (probably largely to libraries rather than individuals) will grow in importance and will be increasingly dominated by a small number of giant corporations. Indeed, we may get a combination of fee-based and advertiser-supported systems. Reed Elsevier's Lexis-Nexis Academic Universe charges substantial subscription fees to libraries but still includes flashing banner ads. (A researcher who is "feeling lucky" can, for example, click a banner and put down some money—perhaps his or her latest research grant—on CybersportsCasino.com's blackjack table.)

To raise an alarm about the capitalist character of the information and publishing business makes little sense since publishing has always been a business. But it has not traditionally been dominated by a few giant corporations. In the fall of 2000 when Reed Elsevier and Thomson jointly purchased the publisher Harcourt (where Ted Nelson thought up the term Xanadu four decades ago) for $4.4 billion in cash and the assumption of $1.2 billion of debt, the *New York Times* observed that the price was below what had been expected. "The main reason for the low price," it explained, "is that consolidation in the educational and professional publishing businesses—Harcourt's core—has progressed so far that there are almost no bidders left. Each of Harcourt's main businesses is dominated by just three or four companies, like McGraw-Hill or Pearson. Almost all potential bidders faced antitrust problems or had balance sheets full from recent acquisitions."[61] In a world in which libraries can only buy from one or two vendors, those vendors can easily dictate prices and content. And in a world in which there are only a few publishers, they can also dictate terms to authors as well.

The advertiser-sponsored online world also seems to be heading down the same path of media consolidation augured by the merger of AOL with Time-Warner, Inc. Consider, for example, the history of *Civil War Times* magazine, whose humble origins go back to the 1940s when LeRoy Smith used his army poker winnings to start some history tourism businesses in Gettysburg, Pennsylvania. In 1962, during the Civil War centennial, he and the newspaperman Robert H. Fowler started *Civil War Times*; later they gradually added some other related history publications to what they called Historical Times, Inc. In 1986, Cowles Media purchased Historical Times, Inc., and added still more history magazines, which became part of "Cowles Enthusiast Media" and the basis of The HistoryNet.com, which appeared on the Web in 1996. Two years later, the McClatchy newspaper chain acquired Cowles and then sold off Cowles Enthusiast Media to Primedia—formerly known as K-III Communications, a conglomerate of specialty magazines (for example, *National Hog Farmer* and *Lowrider Bicycle*) pulled together by the leveraged buyout specialists Kohlberg Kravis Roberts back in the go-go 1980s. In fall 2000, Primedia announced plans to purchase About.com for more than half a billion dollars—thereby not only consolidating old media (magazines) and new (Web) but also bringing together under one corporate umbrella two of the main advertiser-sponsored history sites on the Web. A few months later, it purchased half ownership of Brill Media Holdings, the company behind Contentville.com.[62]

Ironically, despite the trend toward online consolidation, one of the greatest frustrations of the historical Xanadu as it exists at the dawn of the new millennium is its myriad divisions. To find what the Internet offers on Eugene V. Debs requires at least a dozen different searches—through a general search engine such as Google; the scholarly article archives at JSTOR, ProQuest, EAA, EBSCO, the History Cooperative, and Project Muse; reference works at the History Resource Center; the popular history writings at The HistoryNet.com; articles and sources at Contentville; the primary sources at American Memory; and the image archive at Corbis.com. The capitalist market in information and the limitations of Web search engines have fostered both consolidation and competition. Neither trend is wholly friendly to researchers.

Perhaps paradoxically, then, the Web seems to be fostering two contradictory developments. On the one hand, the resources required to publish on the Web are so modest that we have seen an amazing grassroots publishing effort over the past five years. Yet, on the other hand, the capacity to mount a serious Web-based publishing or information business may be quite limited indeed. Even the Web startups such as Questia and NetLibrary are backed by hundreds of millions of dollars in venture capital. To be sure, the nonprofit world also has its giants such as NDLP, but their continuation rests on the shaky base of public sector funding. And Internet-based economies of scale are pushing growing consolidation on a global basis. Will the public History Web survive the onslaught of these mega-operations? Will "authority" and "authenticity" reside with the corporate purveyors of the past? And will corporate vendors find scholarly fastidiousness about accuracy and contextualization as appealing as archivists and academics do?

Bell & Howell president James P. Roemer presents his company—notes *Forbes* magazine—"as the guardian of truth in an Internet free-for-all." "There's no guarantee that what you're getting on the Internet is correct or the information you want," he says. The company spokesman Ben Mondloch puts the significance of its Digital Vault Initiative in even broader terms. "We're the only company that could do this," he told a reporter for *Wired News.* "We've become the de facto nation's archive."[63]

The notion of a privatized and corporatized "national archive" occupies the other end of the continuum from the free and open Xanadu envisioned by Ted Nelson. For a humorous and harrowing glimpse of what that might look like, turn to Neal Stephenson's 1992 cyberpunk novel, *Snow Crash*, in which everything is privately owned, from the FOQNEs (Franchise-Organized Quasi-National Entities) known as Burbclaves, where people live, to the highways run by the competing Fairlanes Inc. and Cruiseways Inc., to the Reverend Wayne's Pearly Gates, which has a monopoly on worship services. The book's protagonist, Hiro Protagonist, is a freelance stringer for the CIC, the Central Intelligence Corporation of Langley, Virginia. The CIC's "database" was, Stephenson writes,

> formerly the Library of Congress, but no one calls it that anymore. Most people are not entirely clear on what the word "congress" means. And even the word "library" is getting hazy. It used to be a place full of books, mostly old ones. Then they began to include videotapes, records, and magazines. Then all of the information got converted into machine-readable form, which is to say, ones and zeroes. And as the number of media grew, the material became more up to date, and the methods for searching the Library became more and more sophisticated, it approached the point where there was no substantive difference between the Library of Congress and the Central Intelligence Agency. Fortuitously, this happened just as the government was falling apart anyway. So they merged and kicked out a big fat stock offering.[64]

It is all too easy in the era of cyberspace to get carried away with extravagant visions of the future—whether the utopian dreams of Ted Nelson or the dystopian vision of *Snow Crash*. History tells us that change comes much more slowly and unevenly than most visionaries would like. Still, what is remarkable is how much the practice of researching, teaching, and presenting the past has changed in the short five years since the Web and Internet entered the lives of historians. We have many reasons to celebrate the enormous advances—the vast archive of primary and secondary sources now accessible on our computer screens and available to us as researchers, to our students, and to anyone concerned about the past. But while we celebrate what has been gained, we should be vigilant about what might be lost if the grassroots energy and the cooperative spirit of enthusiastic amateurs, enterprising librarians, and archivists pursuing personal historical passions and public un-

derstanding of the past are squashed by the advance of a corporate jugger-
naut chasing private profit.

Nevertheless, the power and wealth of the corporate forces should not
lead us to assume that we are headed inevitably toward Stephenson's CIC.
William Y. Arms, the editor of *D-Lib Magazine*, which focuses on digital libraries, has recently argued that "open access" may, in the end, turn out to
dominate the future of information. He observes that whereas 10 years ago
the percentage of information used in professional work that "was available
openly, without payment" was probably one percent or less, today most people would say that 5 to 80 percent is available with open access. I can often
find historical information more quickly on the public Web (and am thus
more likely to use it) than by searching the gated private Web databases that
my university provides to me. My library, for example, pays a thousand dollars a year to get the online version of *Books in Print* from the Thomson Corporation, but Amazon.com provides much of the same information for free.
Increased computer power, moreover, means that it is increasingly easy to
find that information on the vast stretches of the Internet. For Arms, "automated digital libraries combined with open access information on the Internet offer to provide the Model T Ford of information," basic transportation
for all.[65]

Historians have a great stake in shaping the roads and cars that will populate the future information superhighways. We need to put our energies into
maintaining and enlarging the astonishingly rich public historical Web that
has emerged in the past five years. For some, that should mean joining in
eclectic but widespread grassroots efforts to put the past online—whether
that involves posting a few documents online for your students or raising
funds for more ambitious projects to create free public archives. Just as "open
source" code has been the banner of academic computer scientists, "open
sources" should be the slogan of academic and popular historians. Academics
and enthusiasts created the Web; we should not quickly or quietly cede it to
giant corporations. For all of us, shaping the digital future requires a range of
political actions—fighting against efforts to slash the budgets of public agencies such as NEH and the Library of Congress that are funding important
digital projects; resisting efforts further to narrow the "public domain"; and
joining with librarians who have been often alone in raising red flags about
the growing power of the information conglomerates.[66] We may also need to
reexamine our own contradictory position as both rights holders and con-

sumers of copyright content. Perhaps we should even insist that the intellectual property we create (often with considerable public funding) should be freely available to all. Unless we act, the digital Xanadu, as Nelson fantasized, may turn out to have everything an "absent-minded professor could want" but only at and for a heavy price.[67]

W HEN HE ACCEPTED THE RICHARD W. LYMAN
AWARD in May 2003, Roy wanted to recognize all those
with whom he'd collaborated on digital history projects. Because the list he compiled grew to more than 150 names, he knew that reading it aloud would take more time than he'd been allotted for his remarks. So
he put the names on a slide and showed them to his audience. It was a typical
gesture: he derived great pleasure from acknowledging his co-workers; the
readers of his drafts; those who inspired, advised, and encouraged him; and
the schools, agencies, and foundations that provided his work with material
support. I regret that I cannot do justice to what would have been Roy's own
fulsome account of those who accompanied him and enriched his work during the years he was writing about digital history.

He greatly appreciated all the friends and colleagues with whom he created projects at the Center for History and New Media and at the American
Social History Project/Center for Media and Learning. In addition, he and
his co-writers thanked the following people in the articles on new media reproduced in this volume and in *Digital History: A Guide to Gathering, Preserving, and Presenting the Past on the Web*: Debbie Abilock, Mike Alcoff, Susan
Armeny, Emily Bliss, Marta Brooks, Robert Chazan, Rustin Crandall, Leani

Donlan, John Elfrank-Dana, Bret Eynon, Joan Fragazy, Mary Jane Gormley, Abbie Grotke, Giles Hudson, Stephanie Hurter, Kathy Isaacs, Frances Jacobson, Dawn Jaeger, Peter Jaszi, T. Mills Kelly, Bob Lockhart, Joanne Meyerowitz, Rikk Mulligan, Jim O'Brien, Noreen O'Connor, Julie Plaut, Elena Razlogova, Terence Ross, Jim Safley, Carl Schulkin, David Seaman, Peter Seixas, Amanda Shuman, Ron Stoloff, John Summers, Tom Thurston, Bill Tally, David Thelen, Rebecca Tushnet, Eileen Walsh, and John Willinsky. Roy was indebted to the Charles Warren Center at Harvard University for providing the office in which he wrote most of "Scarcity or Abundance?" and to Pat Denault and Laura Thatcher Ulrich for their hospitality there. He thanked the Alfred P. Sloan Foundation for its funding as well as program officer Jesse Ausubel "for his vital encouragement" [and] "visionary commitment to using new digital technology to collect, present, and preserve the past." He also recognized George Mason University's support for the CHNM and particularly the assistance of Daniele Struppa and the steadfast and tireless Jack Censer.

In his new media publications Roy acknowledged some people repeatedly, expressing warm thanks for the insights and wise counsel of Michael Grossberg, Gary Kornblith, Tom Scheinfeldt, Kelly Schrum, Abby Smith, James Sparrow, Robert Townsend, and Sam Wineburg. And for their incisive cultural and political analyses and abundant creativity, he was deeply grateful to those with whom he wrote, edited, and "conspired" in new media for many years: Randy Bass, Stephen Brier, Joshua Brown, Daniel Cohen, and Michael O'Malley.

For their help and advice in the preparation of this book, I wish to thank Jean-Christophe Agnew, Betsy Blackmar, Steve Brier, Josh Brown, Jack Censer, Dan Cohen, Matt Karush, Gary Kornblith, Alison Landsberg, Mike O'Malley, Tom Scheinfeldt, and Sean Takats. Roy never wanted anyone but Jim O'Brien to index his books; I am grateful to him for doing this one as well. It has also been a pleasure to work with Philip Leventhal and Leslie Kriesel, of Columbia University Press. My greatest debt is to Peter Dimock. He first approached Roy with the idea of publishing a collection of his essays on new media, and, with his characteristic tact and gentleness, proposed the plan to me, several months after Roy died. Peter's vision and his faithful persistence made *Clio Wired* possible. Finally, I have dedicated this book to Roy's mother, who taught him—and more recently showed me—how to look forward.

D. K.

Versions of the chapters in this book were previously published in the following venues and are reproduced here with permission:

"Scarcity or Abundance: Preserving the Past in a Digital Era," *American Historical Review* 108, no. 3 (June 2003): 735–762.

"Web of Lies? Historical Knowledge on the Internet," coauthored with Daniel J. Cohen, *First Monday* 10, no. 12 (December 2005).

"Can History Be Open Source? *Wikipedia* and the Future of the Past," *Journal of American History* (June 2006).

"Historians and Hypertext: Is it More Than Hype?" coauthored with Steve Brier, *AHA Perspectives* (March 1994):3–6. (Another version was published in Lawrence Dowler, ed., *Gateways to Knowledge: The Role of Academic Libraries in Teaching, Learning, and Researching* [Cambridge: MIT Press, 1997], 207–214.)

"Rewiring the History and Social Studies Classroom: Needs, Frameworks, Dangers, and Proposals," coauthored with Randy Bass, *Journal of Education* 181, no. 3 (1999), and *Computing in the Social Sciences and Humanities,* ed. Orville Vernon Burton (Urbana: University of Illinois Press, 2002), 20–48.

"The Riches of Hypertext for Scholarly Journals," *Chronicle of Higher Education,* March 17, 2000.

"Should Historical Scholarship Be Free?" *Perspectives on History* 43, no. 4 (2005).

"Collecting History Online" in *Digital History: A Guide to Gathering, Preserving, and Presenting the Past on the Web,* coauthored with Daniel Cohen (Philadelphia: University of Pennsylvania Press, 2005), 160–188.

"Brave New World or Blind Alley? American History on the World Wide Web," coauthored with Michael O'Malley, *Journal of American History* (June 1997):132–155.

"Wizards, Bureaucrats, Warriors, and Hackers: Writing the History of the Internet," *American Historical Review* 103 (December 1998): 1530–52.

"The Road to Xanadu: Public and Private Pathways on the History Web," *Journal of American History* (September 2001):538–579.

1. Scarcity or Abundance? Preserving the Past

1. Greg Miller, "Cyberculture: The Scene/The Webby Awards," *Los Angeles Times*, March 9, 1998, D3. On Ignacio, see the interview "Dino Ignacio: Evil Incarnate," in Philippine Web Designers Network, *Philweavers*, http://www.philweavers.net/ profiles/dinoginacio.html; Buck Wolf, "Osama bin Muppet," *ABC News*, http://www .abcnews.go.com/sections/us/WolfFiles/wolffiles190.html; "Media Killed Bert Is Evil," http://plaza.powersurfr.com/bert/ (viewed online April 15, 2002, but unavailable as of July 4, 2002); Peter Hartlaub, "Bert and bin Laden Poster Tied to S.F. Student," *San Francisco Chronicle*, October 12, 2001, A12; Gina Davidson, "Bert and Bin: How the Joke Went Too Far," *The Scotsman*, October 14, 2001, 3.

2. "Bert Is Evil!" in Snopes.com, http://www.snopes2.com/rumors/bert.htm; "Bert Is Evil—Proof in the Most Unlikely Places," in *HermAphroditeZine*, http:// www.pinktink3.250x.com/hmm/bert.htm; Josh Grossberg, "The Bert-Bin Laden Connection?" in *E! Online News*, October 10, 2001, http://www.eonline.com/News/ Items/0,1,8950,00.html; Joey G. Alarilla, "Infotech Pinoy Webmaster Closes Site After 'Bert-Bin Laden' Link," *Philippine Daily Inquirer,* October 22, 2001, 17; Dino Ignacio, "Good-bye Bert," in *Fractal Cow*, http://www.fractalcow.com/bert/bert.htm. See also Michael Y. Park, "Bin Laden's Felt-Skinned Henchman?" *Fox News* (October 14, 2001), http://www.foxnews.com/story/0,2933,36218,00.html; Declan McCullagh,

"Osama Has a New Friend," *Wired News* (October 10, 2001), http://www.wired.com/news/conflict/0,2100,47450,00.html; "*Sesame Street* Character Depicted with bin Laden on Protest Poster," *AP Worldstream* (October 11, 2001). Nikke Lindqvist, *N!kke*, http://www.lindqvist.com/art.php?incl=bert.php&lang=eng, provides an excellent chronicle of the unfolding story. Significantly, many of the links on this site, which I first viewed in February 2002, were no longer working in March 2003.

3. Jeffrey Benner, "Is U.S. History Becoming History?" *Wired News* (April 9, 2001), http://www.wired.com/news/print/0,1294,42725,00.html.

4. Arcot Rajasekar, Richard Marciano, and Reagan Moore, "Collection-Based Persistent Archives," http://www.sdsc.edu/NARA/Publications/OTHER/Persistent/Persistent.html; U.S. Congress, House Committee on Government Operations, *Taking a Byte out of History: The Archival Presentation of Federal Computer Records*, HR 101–987 (Washington, D.C., 1990); National Academy of Public Administration, *The Effects of Electronic Recordkeeping on the Historical Record of the U.S. Government* (Washington, D.C., 1989), 8, 29; Joel Achenbach, "The Too-Much-Information Age," *Washington Post,* March 12, 1999, A01; General Accounting Office (hereafter, GAO), *Information Management: Challenges in Managing and Preserving Electronic Records* (Washington, D.C., 2002), 11, 66. See also Alexander Stille, *The Future of the Past* (New York, 2002), 306; Richard Harvey Brown and Beth Davis-Brown, "The Making of Memory: The Politics of Archives, Libraries, and Museums in the Construction of National Consciousness," *History of the Human Sciences* 11, no. 4 (1998): 17–32; Deanna Marcum, "*Washington Post* Publishes Letter from Deanna Marcum," *CLIR* (Council on Library and Information Resources) *Issues* no. 2 (March/April 1998), http://www.clir.org/pubs/issues/issues02.html#post.

5. John Higham, *History: Professional Scholarship in America* (1965; rpt., Baltimore, 1983), 16–20. See also American Historical Association Committee on Graduate Education, *The Education of Historians in the 21st Century* (Urbana, Ill., 2004). To observe this broader vision is not to deny the very different historical circumstances (such as the disorganization of archives), the obvious blindness of the early professional historians on many matters (such as race and gender), and the early tensions between "amateurs" and professionals.

6. For interesting observations on "abundance" in two different realms of historical work, see James O'Toole, "Do Not Fold, Spindle, or Mutilate: *Double Fold* and the Assault on Libraries," *American Archivist* 64 (Fall/Winter 2001): 385–93; John McClymer, "Inquiry and Archive in a U.S. Women's History Course," *Works and Days* 16, nos. 1–2 (Spring/Fall 1998): 223. For a sweeping statement about political and cultural implications of "digital information that moves frictionlessly through the network and has zero marginal cost per copy," see Eben Moglen, "Anarchism Triumphant: Free Software and the Death of Copyright," *First Monday* 4, no. 8 (August 1999), http://www.firstmonday.dk/issues/issue4_8/moglen/index.html.

7. Committee on the Records of Government, *Report* (Washington, D.C., 1985), 9 (the committee was created by the American Council of Learned Societies, the Council on Library Resources, and the Social Science Research Council with funding from the Mellon, Rockefeller, and Sloan foundations); John Garrett and Donald Waters, *Preserving Digital Information: Report of the Task Force on Archiving of Digital Information* (Washington, D.C., 1996); Paul Conway, *Preservation in the Digital World* (Washington, D.C., 1996), http://www.clir.org/pubs/reports/conway2/index.html. For other reports with similar conclusions, see, for example, the 1989 report of the National Association of Government Archives and Records Administrators, cited in Margaret Hedstrom, "Understanding Electronic Incunabula: A Framework for Research on Electronic Records," *American Archivist* 54 (Summer 1991): 334–54; House Committee on Government Operations, *Taking a Byte out of History*; Committee on an Information Technology Strategy for the Library of Congress, Computer Science and Telecommunications Board, Commission on Physical Sciences, Mathematics, and Applications, and the National Research Council, *LC21: A Digital Strategy for the Library of Congress* (Washington, D.C., 2000), http://books.nap.edu/html/lc21/index.html; GAO, *Information Management; NHPRC Electronic Records Agenda Final Report (Draft)* (St. Paul, Minn., 2002).

8. Margaret MacLean and Ben H. Davis, eds., *Time and Bits: Managing Digital Continuity* (Los Angeles, 1998), 11, 6; Jeff Rothenberg, *Avoiding Technological Quicksand: Finding a Viable Technical Foundation for Digital Preservation* (Washington, D.C., 1998), http://www.clir.org/pubs/reports/rothenberg/contents.html. The 1997 conference "Documenting the Digital Age" has also disappeared from the Web, nor is it available in the Internet Archive. The Sanders film is available from the Council on Library and Information Resources, http://www.clir.org/pubs/film/future/order.html.

9. Achenbach, "Too-Much-Information Age." See also Stille, *Future of the Past*; Council on Library and Information Resources, *The Evidence in Hand: Report of the Task Force on the Artifact in Library Collections* (Washington, D.C., 2001), http://www.clir.org/pubs/reports/pub103/contents.html.

10. Margaret O. Adams and Thomas E. Brown, "Myths and Realities about the 1960 Census," *Prologue: Quarterly of the National Archives and Records Administration* 32, no. 4 (Winter 2000), http://www.archives.gov/publications/prologue/winter_2000_1960_census.html. See also letter of August 15, 1990, from Kenneth Thibodeau, which says that recovering the records took "substantial efforts" by the Bureau of the Census, quoted in House Committee on Government Operations, *Taking a Byte out of History*, 3. According to Timothy Lenoir, it is now too expensive to rescue the computer tapes that represent Douglas Englebart's pioneering hyperrmedia-groupware system called NLS (for oNLine System)—the basis of many of the features of personal computers. Timothy Lenoir, "Lost in the Digital Dark Ages" (paper delivered at

"The New Web of History: Crafting History of Science Online," Cambridge, Mass., March 28, 2003).

11. Marcia Stepanek, "From Digits to Dust," *Business Week* (April 20, 1998); House Committee on Government Operations, *Taking a Byte out of History*, 16; Jeff Rothenberg, "Ensuring the Longevity of Digital Documents," *Scientific American* (January 1995):42–47. See also Garrett and Waters, *Preserving Digital Information*. Many Vietnam records are stored in a database system that is no longer supported and can only be translated with difficulty. As a result, the Agent Orange Task Force could not use important herbicide records. Stille, *Future of the Past*, 305.

12. Most Microsoft software moves into what the company calls the "nonsupported phase" after just four or five years, although it offers a more limited "extended support phase" that lasts up to seven years. After that, you are out of luck. Microsoft, "Windows Desktop Product Life Cycle Support and Availability Policies for Businesses," October 15, 2002, http://www.microsoft.com/windows/lifecycle.mspx; Lori Moore, "Q&A: Microsoft Standardizes Support Lifecycle," *Press Pass: Information for Journalists* (October 15, 2002), http://www.microsoft.com/presspass/features/2002/Oct02/10-15support.asp. On media longevity, see Rothenberg, *Avoiding Technological Quicksand*; MacLean and Davis, *Time and Bits*; Margaret Hedstrom, "Digital Preservation: A Time Bomb for Digital Libraries" (paper delivered at the NSF Workshop on Data Archiving and Information Preservation, March 26–27, 1999), http://www.uky.edu/~kiernan/DL/hedstrom.html; Frederick J. Stielow, "Archival Theory and the Preservation of Electronic Media: Opportunities and Standards Below the Cutting Edge," *American Archivist* 55 (Spring 1992): 332–43; Charles M. Dollar, *Archival Theory and Information Technology: The Impact of Information Technologies on Archival Principles and Methods* (Ancona, Italy, 1992), 27–32; GAO, *Information Management*, 50–52.

13. Richard J. Cox, "Messrs. Washington, Jefferson, and Gates: Quarrelling About the Preservation of the Documentary Heritage of the United States," *First Monday* 2, no. 8 (August 1997), http://firstmonday.org/issues/issue2_8/cox/. See also Peter Lyman and Brewster Kahle, "Archiving Digital Cultural Artifacts: Organizing an Agenda for Action," *D-Lib Magazine* 4, nos. 7–8 (July/August 1998), http://www.dlib.org/dlib/july98/07lyman.html. Voyager's CD-ROM explicating Beethoven's Ninth Symphony—a landmark work in multimedia—no longer operates, in part because Apple changed a CD-ROM driver that the program relied on. Robert Winter, *Ludwig Van Beethoven Symphony No. 9* (Santa Monica, Calif., 1991). Digital art presents particularly difficult problems; see, for example, Scott Carlson, "Museums Seek New Methods for Preserving Digital Art," *Chronicle of Higher Education* (August 16, 2002).

14. Margaret Hedstrom, "How Do We Make Electronic Archives Usable and Accessible?" (paper delivered at "Documenting the Digital Age," San Francisco, February 10–12, 1997); Luciana Duranti, "Diplomatics: New Uses for an Old Science," *Archivaria* 28 (Summer 1989): 7–27; Peter B. Hirtle, "Archival Authenticity in a Digital Age," in *Council on Library and Information Resources, Authenticity in a Digital Environ-*

ment (Washington, D.C., 2000), http://www.clir.org/pubs/reports/pub92/contents. html; CLIR, *Evidence in Hand*; Susan Stellin, "Google's Revival of a Usenet Archive Opens Up a Wealth of Possibilities But Also Raises Some Privacy Issues," *New York Times*, May 7, 2001, C4; David Bearman and Jennifer Trant, "Authenticity of Digital Resources: Towards a Statement of Requirements in the Research Process," *D-Lib Magazine* 4, no. 6 (June 1998), http://www.dlib.org/dlib/june98/06bearman.html.

15. Abby Smith, "Authenticity in Perspective," in CLIR, *Authenticity in a Digital Environment* (Washington, D.C., 2000), http://www.clir.org/pubs/reports/pub92/ smith.html; Clifford Lynch, "Authenticity and Integrity in the Digital Environment: An Exploratory Analysis of the Central Role of Trust," in *Authenticity in a Digital Environment*, http://www.clir.org/pubs/reports/pub92/contents.html. See also M. T. Clanchy, *From Memory to Written Record: England 1066–1307*, 2d ed. (Oxford, 1993); see also Research Libraries Group, *Attributes of a Trusted Digital Repository: Meeting the Needs of Research Resources; An RLC-OCLC Report* (Mountain View, Calif., 2001), http://www.rlg.org/longterm/attributes01.pdf.

16. Brewster Kahle, Rick Prelinger, and Mary E. Jackson, "Public Access to Digital Materials" (white paper delivered at the Association of Research Libraries and Internet Archive Colloquium "Research in the 'Born-Digital' Domain," San Francisco, March 4, 2001), http://www. dlib.org/dlib/october01/kahle/10kahle.html.

17. Committee on Intellectual Property Rights in the Emerging Information Infrastructure, National Research Council, et al., *The Digital Dilemma: Intellectual Property in the Information Age* (Washington, D.C., 1999), http://books.nap.edu/html/digital_ dilemma/; Richard Stallman, "Can You Trust Your Computer?" *Newsforge* (October 21, 2002), http://newsforge.com/newsforge/02/10/21/1449250.shtml?tid=19. The Digital Millennium Copyright Act makes it illegal to circumvent technical protection services. See Peter Lyman, "Archiving the World Wide Web," in CLIR, *Building a National Strategy for Digital Preservation: Issues in Digital Media Archiving* (Washington, D.C., 2002), http://www.clir.org/pubs/reports/pub106/web.html.

18. CLIR, *Evidence in Hand*.

19. Committee on Intellectual Property Rights in the Emerging Information Infrastructure, National Research Council, *Digital Dilemma*.

20. As with our network of research libraries, this system is a modern invention. The first public governmental archive came with the French Revolution; the British Public Record Office opened in 1838, and the National Archives is of startlingly recent vintage: the legislation establishing it did not come until 1934. Donald R. McCoy, "The Struggle to Establish a National Archives in the United States," in *Guardian of Heritage: Essays on the History of the National Archives*, ed. Timothy Walch (Washington, D.C., 1985), 1–15.

21. Don Waters, "Wrap Up" (paper delivered at the DAI Institute, "The State of Digital Preservation: An International Perspective," Washington, D.C., April 25, 2002), http://www.clir.org/pubs/reports/pub107/contents.html; Dale Flecker, "Pre-

serving Digital Periodicals," in *CLIR, Building a National Strategy for Digital Preservation*.

22. Michael L. Miller, "Assessing the Need: What Information and Activities Should We Preserve?" (paper delivered at "Documenting the Digital Age," San Francisco, February 10–12, 1997). To be sure, it has been biased toward the preservation of the records of the rich and powerful, although in more recent years energetic "activist archivists" have sought out more diverse sets of materials. Ian Johnston, "Whose History Is It Anyway?" *Journal of the Society of Archivists* 22, no. 2 (2001): 213–29.

23. See Adrian Cunningham, "Waiting for the Ghost Train: Strategies for Managing Electronic Personal Records Before It Is Too Late" (paper delivered at the Society of American Archivists Annual Meeting, Pittsburgh, August 23–29, 1999), http://www.rbarry.com/cunningham-waiting2.htm. For numbers of commercial word-processing programs, see House Committee on Government Operations, *Taking a Byte out of History*, 15.

24. *SRA International, Report on Current Recordkeeping Practices within the Federal Government* (Arlington, Va., 2001), http://www.archives.gov/records_management/pdf/report_on_recordkeeping_practices.pdf. This report responded to an earlier GAO report: U.S. Government Accounting Office, National Archives: *Preserving Electronic Records in an Era of Rapidly Changing Technology* (Washington, D.C., 1999). Archival consultant Rick Barry reports that four-fifths of e-mail creators he surveyed "do not have a clue" whether their e-mail was an official record and that most are "largely unaware" of official e-mail policies. Quoted in David A. Wallace, "Recordkeeping and Electronic Mail Policy: The State of Thought and the State of the Practice" (paper delivered at the Annual Meeting of the Society of American Archivists, Orlando, Florida, September 3, 1998), http://www.rbarry.com/wallace.html.

25. Rothenberg, *Avoiding Technological Quicksand*. For the long controversy over NARA and the printing of e-mail, see Bill Miller, "Court Backs Archivist's Rule: U.S. Agencies May Be Allowed to Delete E-Mail," *Washington Post,* August 7, 1999, A02; Wallace, "Recordkeeping and Electronic Mail Policy"; GAO, *Information Management*, 57–65.

26. See Stewart Granger, "Emulation as a Digital Preservation Strategy," *D-Lib Magazine* 6, no. 10 (October 2000), http://www.dlib.org/dlib/october00/granger/10granger.html, on this as the "dominant" approach. An even earlier intervention version of "migration" is to move digital objects to "standardized" formats immediately or as quickly as possible, to put them in nonproprietary, open-source, commonly accepted formats (for instance, ASCII for text, .tiff for images, etc.) that are likely to be around for a long time. Of course, popular standards are no guarantee of longevity; in 1990, NARA was arguing that spreadsheets formatted for Lotus 1-2-3 were not a preservation problem since the program was so "widespread." House Committee on Government Operations, *Taking a Byte out of History*, 12.

27. Warwick Cathro, Colin Webb, and Julie Whiting, "Archiving the Web: The PANDORA Archive at the National Library of Australia" (paper delivered at "Preserving the Present for the Future Web Archiving," Copenhagen, June 18–19, 2001). See also Diane Vogt-O'Connor, "Is the Record of the 20th Century at Risk?" *CRM: Cultural Resource Management* 22, no. 2 (1999): 21–24.

28. Rothenberg, *Avoiding Technological Quicksand.*

29. Margaret Hedstrom, "Digital Preservation: Matching Problems, Requirements and Solutions" (paper delivered at the NSF Workshop on Data Archiving and Information Preservation, March 26–27, 1999), http://cecssrv1.cecs.missouri.edu/NSFWorkshop/hedpp.html (accessed March 2002 but unavailable in May 2003). See also Margaret Hedstrom, "Research Issues in Digital Archiving" (paper delivered at the DAI Institute, "The State of Digital Preservation: An International Perspective, Washington, D.C., April 25, 2002, http://www.clir.org/pubs/reports/pub107/contents.html). Rothenberg himself is currently undertaking research on emulation, and other emulation research is going on at the University of Michigan and Leeds University and at IBM's Almaden Research Center in San Jose, California. Daniel Greenstein and Abby Smith, "Digital Preservation in the United States: Survey of Current Research, Practice, and Common Understandings" (paper delivered at "Preserving History on the Web: Ensuring Long-Term Access to Web-Based Documents," Washington, D.C., April 23, 2002). More recently, Rothenberg has apparently tempered his position on emulation versus migration.

30. David Bearman and Jennifer Trant, "Electronic Records Research Working Meeting, May 28–30, 1997: A Report from the Archives Community," *D-Lib Magazine* 3, nos. 7–8 (July/August 1997), http://www.dlib.org/dlib/july97/07bearman.html; Terry Cook, "The Impact of David Bearman on Modern Archival Thinking: An Essay of Personal Reflection and Critique," *Archives and Museum Informatics* 11 (1997): 23. See further Margaret Hedstrom, "Building Record-Keeping Systems: Archivists Are Not Alone on the Wild Frontier," *Archivaria* 44 (Fall 1997): 46–48. See also David Bearman and Ken Sochats, "Metadata Requirements for Evidence," in University of Pittsburgh, School of Information Sciences, the Pittsburgh Project, http://www.archimuse.com/papers/NHPRC/. (Many parts of this site have disappeared, but this undated paper is available at http://www.archimuse.com/papers/NHPRC/BACartic.html.) David Bearman, "An Indefensible Bastion: Archives as Repositories in the Electronic Age," in Bearman, ed., *Archival Management of Electronic Records* (Pittsburgh, 1991), 14–24; Margaret Hedstrom, "Archives as Repositories—A Commentary," in ibid.

31. Cook, "Impact of David Bearman on Modern Archival Thinking," 15–37. From another perspective, the Pitt Project broadened, rather than narrowed, the concerns of electronic archivists, since previously the focus had been on statistical databases. In one effort to join the emphasis on records as evidence with a broader social cultural focus, Margaret Hedstrom argues that "to benefit fully from the synergy between

business needs and preservation requirements, cultural heritage concerns should be linked to equally critical social goals, such as monitoring global environment change, locating nuclear waste sites, and establishing property rights, all of which also depend on long-term access to reliable, electronic evidence." Quoted in Richard J. Cox, "Searching for Authority: Archivists and Electronic Records in the New World at the Fin-de-Siècle," *First Monday* 5, no. 1 (January 3, 2000), http://firstmonday.org/issues/ issue5_1/cox/index.html. The Pitt Project has been the subject of enormous discussion and significant debate among archivists; a full and nuanced treatment of the subject is beyond the scope of this article. Whereas Cook offers serious criticism of Bearman, the leader of the project along with Richard Cox, he also celebrates Bearman as "the leading archival thinker of the late twentieth century." Linda Henry offers a sweeping attack on Bearman and other advocates of a "new paradigm" in electronic records management in "Schellenberg in Cyberspace," *American Archivist* 61 (Fall 1998): 309–27. A more recent critique is Mark A. Greene, "The Power of Meaning: The Archival Mission in the Postmodern Age," *American Archivist* 65, no. 1 (Spring/Summer 2002): 42–55. Terry Cook puts the story in historical perspective (but from his particular perspective) in "What Is Past Is Prologue: A History of Archival Ideas Since 1898, and the Future Paradigm Shift," *Archivaria* 43 (Spring 1997), http://www.rbarry.com/cookt-pastprologue-ar43fnl.htm. The project "Preservation of the Integrity of Electronic Records" (called the UBC Project because it was carried out at the University of British Columbia) and the InterPARES project (International Research on Permanent Authentic Records in Electronic Systems), which built on the UBC Project, have taken a different approach, but they share the Pitt Project's emphasis on the problem of "authenticity" and on "records" rather than the broader array of sources that generally interest historians. Luciana Duranti, *The Long-Term Preservation of Authentic Electronic Records: Findings of the InterPARES Project* (Vancouver, 2002), http://www.interpares.org/book/index.htm. The December 2002 draft of the NHPRC Electronic Records Agenda Final Report suggests that the consensus among archivists is moving toward a broader definition of records. My understanding of these issues has been greatly aided by attending the December 8–9, 2002, meeting convened to discuss that agenda and by conversations with Robert Horton of the Minnesota Historical Society, who is the leader of that effort.

32. Carolyn Said, "Archiving the Internet: Brewster Kahle Makes Digital Snapshots of Web," *San Francisco Chronicle,* May 7, 1998, B3; Brewster Kahle, "Preserving the Internet," *Scientific American* (March 1997), http://www.sciamdigital.com; Kendra Mayfield, "Wayback Goes Way Back on Web," *Wired News* (October 29, 2001), http:// www.wired.com/news/print/0,1294,47894,00.html; Mike Burner, "The Internet Archive Robot," e-mail to Robots Mailing List, September 5, 1996, http://www.robot stxt.org/wc/mailing-list/1258.html. On Alexa, see Rajiv Chandrasekaran, "Seeing the Sites on a Custom Tour: New Internet Search Tool Takes Selective Approach," *Washington Post,* September 4, 1997, E01; Tim Jackson, "Archive Holds Wealth of Data,"

Financial Times (London), November 24, 1997, 15; Laurie J. Flynn, "Alexa's Crusade Continues Under Amazon.com's Flag," *New York Times,* May 3, 1999, C4. On other early efforts to "save the Web," see Spencer Reiss, "Internet in a Box," *Wired* (October 1996), http://www.wired.com/wired/4.10/scans.html; Bruce Sterling, "The Life and Death of Media" (speech delivered at the Sixth International Symposium on Electronic Art, Montreal, September 19, 1995), http://www.chriswaltrip.com/sterling/dedmed.html; John Markoff, "When Big Brother Is a Librarian," *New York Times,* March 9, 1997, IV:3; James B. Gardner, comp., "Report on Documenting the Digital Age" (Washington, D.C., 1997); Nathan Myhrvold, "Capturing History Digitally: Why Archive the Internet?" (paper delivered at "Documenting the Digital Age," San Francisco, February 10–12, 1997).

33. Hamish Mackintosh, Interview with Brewster Kahle, "Webarian," *Guardian,* February 21, 2002, 4, http://www.guardian.co.uk/online/story/0,3605,653286,00.html; Molly Wood, "CNET's Web Know-It-All Goes Where You Won't," *CNET* (March 15, 2002), http://www.cnet.com/software/0–8888–8-9076625–1.html; "Seeing the Future in the Web's Past," *BBC News* (November 12, 2001), http://news.bbc.co.uk/hi/english/in_depth/sci_tech/2000/dot_life/newsid_1651000/1651557.stm. For a good explanation of the technical side of IA, see Richard Koman, "How the Wayback Machine Works," *O'Reilly Network* (January 21, 2002), http://www.oreillynet.com/pub/a/webservices/2002/01/18/brewster.html.

34. Google Employee, "Google Groups Archive Information Newsgroups," e-mail, December 21, 2001; Stellin, "Google's Revival of a Usenet Archive Opens Up a Wealth of Possibilities"; Danny Fortson, "Google Gobbles Up Deja.com's Babble," *Daily Deal* (February 12, 2001); Michael Liedtke, "Web Search Engine Google Buys Deja.com's Usenet Discussion Archives," *Associated Press* (February 12, 2001).

35. Lyman, "Archiving the World Wide Web."

36. Miller quoted in Gardner, "Report on Documenting the Digital Age." For an overview of OAIS, see Brian Lavoie, "Meeting the Challenges of Digital Preservation: The OAIS Reference Model," 2000, http://www.oclc.org/research/publications/newsletter/repubs/lavoie243/; on EAD, see Daniel V. Pitti, "Encoded Archival Description: An Introduction and Overview," *D-Lib Magazine* 5, no. 11 (November 1999), http://www.dlib.org/dlib/november99/11pitti.html. OAIS comes out of NASA and the space data community, not the librarians. But they have embraced it.

37. Raymie Stata, "The Internet Archive" (paper delivered at the conference "Preserving Web-Based Documents," Washington, D.C., April 23, 2002). On deep versus surface Web, see Lyman, "Archiving the World Wide Web"; "The Road to Xanadu: Public and Private Pathways on the History Web," in this volume. Kahle himself indicates many of the problems and limitations of the Internet Archive in Brewster Kahle, "Archiving the Internet: Bold Efforts to Record the Entire Internet Are Expected to Lead to New Services" (paper presented at "Documenting the Digital Age," San Francisco, February 10–12, 1997).

38. On robots exclusion, see http://www. robotstxt.org/wc/exclusion-admin.html. Apparently, the IA will retroactively block a site without direct request, if it simply posts the robots.txt file. This would seem to mean that if someone took over an expired domain name, they could then block access to the prior content. There is some evidence, however, that the IA does not actually "purge" the content, it simply makes it inaccessible. For an intense discussion of these issues, see the hundreds of online postings in "The Wayback Machine, Friend or Foe?" *Slashdot* (June 19–20, 2002), http://ask.slashdot.org/askslashdot/02/06/19/1744209.shtml. For a pessimistic assessment of the legality of the IA's practices (though not explicitly directed at it), see I. Trotter Hardy, "Internet Archives and Copyright" (paper delivered at "Documenting the Digital Age," San Francisco, February 10–12, 1997).

39. Insiders have commented to me that the IA would disappear if Kahle left the project. But there are very recent signs that the IA is broadening its base of financial support.

40. For a recent, brief overview of these trends, see Naomi Klein, "Don't Fence Us In," *Guardian* (October 5, 2002).

41. Thomas Brown, "What Is Past Is Analog: The National Archives Electronic Records Program Since 1968" (paper delivered at the OAH Annual Meeting, Washington, D.C., 2002). In 1997, Kenneth Thibodeau estimated that the NARA invested only token amounts (2 percent of its budget) in electronic records. Gardner, "Report on Documenting the Digital Age."

42. Committee on an Information Technology Strategy for the Library of Congress, et al., LC21, http://books.nap.edu/html/lc21/index.html; "The Road to Xanadu."

43. "Background Information About PANDORA: The National Collection of Australian Online Publications," PANDORA, http://pandora.nla.gov.au/background. html; Cathro, Webb, and Whiting, "Archiving the Web"; Colin Webb, "National Library of Australia" (paper delivered at the DAI Institute, "The State of Digital Preservation: An International Perspective," Washington, D.C., April 25, 2002, http://www. clir.org/pubs/reports/pub107/contents.html). For British efforts to cope with digital materials, see Jim McCue, "Can You Archive the Net?" *Times* (London), April 29, 2002. On Sweden and Norway, see Warwick Cathro, "Archiving the Web," *National Library of Australia Gateways* 52 (August 2001), http://www.nla.gov.au/ntwkpubs/ gw/52/p11a01.html/.

44. There is anecdotal evidence that this is being seriously considered.

45. National Archives and Records Administration, *Proposal for a Redesign of Federal Records Management* (July 2002), 10, http://www.archives.gov/records_manage ment/initiatives/ rm_redesign.html; Richard W. Walker, "For the Record, NARA Techie Aims to Preserve," *Government Computer News* 20, no. 21 (July 30, 2001), http:// www.gcn.com/vol20_no21/news/4752–1.html/; GAO, *Information Management*, 50. So far, POP remains, as a NARA staff member explained in April 2001, "beyond the

state of the art of information technology." Adrienne M. Woods, "Toward Building the Archives of the Future" (paper delivered at the Society of California Archivists' Annual Meeting, April 27, 2001), accessed online May 1, 2002, but not available as of June 20, 2002. See also Kenneth Thibodeau, "Overview of Technological Approaches to Digital Preservation and Challenges in Coming Years" (presentation at the DAI Institute, "The State of Digital Preservation: An International Perspective," Washington, D.C., April 24–25, 2002, http://www.clir.org/pubs/reports/pub107/contents. html). In June 2002, the GAO reported that, in general, NARA's electronic records project "faces substantial risks" and "is already behind schedule." GAO, *Information Management*, 3.

46. Amy Friedlander, "The National Digital Information Infrastructure Preservation Program: Expectations, Realities, Choices and Progress to Date," *D-Lib Magazine* 8, no. 4 (April 2002), http://www.dlib.org/dlib/april02/friedlander/04friedlander. html.

47. The quote is often incorrectly attributed to Carl von Clausewitz. It could be that it is simply a reworking of Voltaire's remark that "le mieux est l'enemi du bien" (the best is the enemy of the good) or of George S. Patton's dictum, "A good plan violently executed now is better than a perfect plan executed next week."

48. Kahle, Prelinger, and Jackson, "Public Access to Digital Materials." See, similarly, Michael Lesk, "How Much Information Is There in the World?" an online paper at http://www.lesk.com/mlesk/ksg97/ksg.html.

49. McCoy, "Struggle to Establish a National Archives in the United States," 1, 12. Indeed, one digital preservation program—LOCKSS (Lots of Copies Keep Stuff Safe)—relies on precisely this principle: http://lockss.stanford.edu/.

50. Lee Dembart, "Go Wayback," *International Herald Tribune* (March 4, 2002), http://www.iht.com/cgi-bin/generic.cgi?template=articleprint.tmplh&ArticleId= 50002; "Seeing the Future in the Web's Past," *BBC News* (November 12, 2001). See also Joseph Menn, "Net Archive Turns Back 10 Billion Pages of Time," *Los Angeles Times,* October 25, 2001, A1; Heather Green, "A Library as Big as the World," *Business Week Online* (February 28, 2002), http://www.businessweek.com/technology/content/feb2002/tc20020228_1080.htm. The dream of a universal archive is also the nightmare of privacy advocates. In the paper era, the physical bulk of personnel files and bank, criminal, and medical records made them more likely to wind up in landfills than in archives. Even when preserved, the possibility of retrospective prying (was your neighbor's grandfather a deadbeat or a drunk?) was reduced by the sheer tedium of sorting through thousands of pages of records. But what if sophisticated data-mining tools ("tell me everything about my neighbors") made such searching easy? Even the "public" material on the Web poses ethical challenges for historians. "The woman who is going to be elected president in 2024 is in high school now, and I bet she has a home page," exclaims Kahle. The Internet Archives has "the future president's home page!" Perhaps. But it also has the home pages of many other high school students, at

least some of whom are going through serious emotional turmoil that they might later prefer to keep from public view. Kahle himself wrote a prescient 1992 article, "Ethics of Digital Librarianship," which worries about "types of information that will be accessible" as "the system grows to include entertainment, employment, health and other servers." Menn, "Net Archive Turns Back 10 Billion Pages"; Wood, "CNET's Web Know-It-All"; Kahle quoted in John Markoff, "Bitter Debate on Privacy Divides Two Experts," *New York Times,* December 30, 1999, C1. See also Jean-François Blanchette and Deborah G. Johnson, "Data Retention and the Panoptic Society: The Social Benefits of Forgetfulness," *Information Society* 18 (2002): 33–45; Marc Rotenberg, "Privacy and the Digital Archive: Outlining Key Issues" (paper delivered at "Documenting the Digital Age," San Francisco, February 10–12, 1997); "Wayback Machine, Friend or Foe?"

51. Mike Featherstone, "Archiving Cultures," *British Journal of Sociology* 51, no. 1 (January 2000): 178, 166. For examples of enthusiastic prophecy about such changes, see Francis Cairncross, *The Death of Distance: How the Communications Revolution Will Change Our Lives* (Boston, 1997); Kevin Kelly, "New Rules for the New Economy," *Wired* 5, no. 9 (September 1997), http://www.wired.com/wired/archive/5.09/new rules_pr.html. For a sober and sensible critique, see John Seely Brown and Paul Duguid, *The Social Life of Information* (Boston, 2000), 11–33.

52. See, for example, Geoffrey J. Giles, "Archives and Historians: An Introduction," in *Archives and Historians: The Crucial Partnership* (Washington, D.C., 1996), 5–13, who writes that "there is too much archival material for the archivists and for the historian to deal with" and notes feelings of "envy" of "ancient and medieval historians, who have so little material with which to work."

53. A. R. Luria, *The Mind of a Mnemonist: A Little Book About a Vast Memory*, trans. Lynn Solotaroff (New York, 1968), foreword by Jerome S. Bruner, viii. See the similar, but fictional, account in Jorge Luis Borges, "Funes the Memorious," in *Labyrinths: Selected Stories and Other Writings*, eds. Donald A. Yates and James E. Irby (New York, 1964), 59–66.

54. Linton Weeks, "Power Biographer," *Washington Post,* April 25, 2002, C01. Carl Bridenbaugh's derisive view of sampling provides a good example of the traditional view that historians should look at everything. "The Great Mutation," *AHR* 68, no. 2 (January 1963): 315–31, also available with other Presidential Addresses at http://www.theaha.org/info/AHA_History/cbridenbaugh.htm. Nevertheless, historians have always struggled with the problem of how to deal with large numbers of sources. Even medievalists worry about how to make sense of the huge numbers of documents that survive from twelfth-century Italy. Still, the digital era vastly increases the scale of the problem.

55. Stellin, "Google's Revival of a Usenet Archive Opens Up a Wealth of Possibilities"; Hedstrom, "How Do We Make Electronic Archives Usable and Accessible?"

(paper delivered at "Documenting the Digital Age," San Francisco, February 10–12, 1997).

56. To be sure, a number of key figures in digital archives and library circles (for example, Daniel Greenstein, Margaret Hedstrom, Abby Smith, Kenneth Thibodeau, Bruce Ambacher) have doctoral degrees in history, but they do not currently work as academic historians. Still, it would be logical for academic historians to build alliances with these scholars who have a foot in both camps. Thus far, academic historians have been much more likely to build ties to historians working in museums and historical societies than to those in archives and libraries.

57. It is difficult to prove a negative, but one searches in vain through the participant lists at key digital archives conferences for the names of practicing historians. One exception was the Committee on the Records of Government, which had a historian, Ernest R. May, as its chair and another, Anna K. Nelson, as its project director. But perhaps significantly, that committee had a mandate that dealt as much with paper as electronic records: Committee on the Records of Government, *Report* (1985). Another partial exception was the February 1997 conference "Documenting the Digital Age" sponsored by NSF, MCI Communications Corporation, Microsoft Corporation, and History Associates Incorporated, which included a few public and museum-based historians but only one university-based historian. Similarly, history journals have provided almost no coverage of these issues. Archivists are not reading historians, either. Richard Cox analyzed the almost 1,200 citations in 61 articles on electronic records management published in the 1990s and found only a handful of references to work by historians. Cox, "Searching for Authority."

58. Cox, "Messrs. Washington, Jefferson, and Gates." Robert Townsend, Assistant Director of Research and Publications, AHA, kindly supplied membership information. One imperfect but telling indicator of the changing interests of professional historians: between 1895 and 1999, the *American Historical Review* published thirty-one articles with one of the following words in the title: archive or archives, records, manuscripts, correspondence. Only four of those appeared after World War II, and they were in 1949, 1950, 1952, and 1965. Some representative titles include: Charles H. Haskins, "The Vatican Archives," *AHR* 2, no. 1 (October 1896): 40–58; Waldo Gifford Leland, "The National Archives: A Programme," *AHR* 18, no. 1 (October 1912): 1–28; Edward G. Campbell, "The National Archives Faces the Future," *AHR* 49, no. 3 (April 1944): 441–45. For a good, brief overview of the AHA's active, early archive and manuscript work, see Arthur S. Link, "The American Historical Association, 1884–1984: Retrospect and Prospect," *AHR* 90, no. 1 (February 1985): 1–17. NARA's "Timeline for the National Archives and Records Administration and the Development of the U.S. Archival Profession," http://www.archives.gov/research_room/alic/reference_desk/NARA_timeline.html, highlights the role of the AHA. It should be noted, however, that the AHA has made a notable contribution to archival issues

through its central role in the National Coordinating Committee for the Promotion of History (NCC), which was crucial, for example, in winning the independence of the National Archives in 1984. The new National Coalition for History, which has replaced the NCC, has also made archival concerns central to its work. Access to archives and primary sources was, of course, a central preoccupation—indeed, an obsession—of early "scientific" and professional historians. See Bonnie G. Smith, "Gender and the Practices of Scientific History: The Seminar and Archival Research in the Nineteenth Century," *AHR* 100, no. 4 (October 1995): 1150–76.

59. Deanna B. Marcum, "Scholars as Partners in Digital Preservation," *CLIR Issues*, no. 20 (March/April 2001), http://www.clir.org/pubs/issues/issues20.html. "Scholars," warns the CLIR Task Force on the Artifact in Library Collections, "may not see preservation of research collections as their responsibility, but until they do, there is a risk that many valuable research sources will not be preserved." CLIR, *Evidence in Hand*.

60. I am indebted to Jim Sparrow for a number of the ideas in this paragraph. For detailed coverage of "How the Bellesiles Story Developed," see *History News Network*, http://hnn.us/articles/691.html.

61. House Committee on Government Operations, *Taking a Byte out of History*, 4. For the assumption of selectivity among archivists, see, for instance, Richard J. Cox, "The Great Newspaper Caper: Backlash in the Digital Age," *First Monday* 5, no. 12 (December 2000), http://firstmonday.org/issues/issue5_12/cox/index.html.

62. Abby Smith, *The Future of the Past: Preservation in American Research Libraries* (Washington, D.C., 1999), www.clir.org/pubs/reports/pub82/pub82text.html; Marcum, "Scholars as Partners in Digital Preservation"; Nicholson Baker, *Double Fold: Libraries and the Assault on Paper* (New York, 2001). Compare, for example, Cox, "Great Newspaper Caper," and O'Toole, "Do Not Fold, Spindle, or Mutilate," with Robert Darnton, "The Great Book Massacre," *New York Review of Books* (April 26, 2001), www.nybooks.com/articles/14196. In 1996, the Modern Language Association (MLA) issued a statement arguing "that for practical purposes, all historical publications, even those produced by mass-production techniques designed to minimize deviations from a norm, have unique physical qualities that may have value as a carrier of (physical) evidence in a given research project." CLIR, *Evidence in Hand*.

63. GAO, *Information Management*, 16; Cox, "Messrs. Washington, Jefferson, and Gates." Cox's article responded, in part, to an earlier article by Raymond W. Smock that argues, "historians should not rely on archivists alone to make decisions about what history to save or to publish." Smock, "The Nation's Patrimony Should Not Be Sacrificed to Electronic Records," *Chronicle of Higher Education,* February 14, 1997, B4–5.

64. Robert Darnton, Sarah A. Mikel, and Shirley K. Baker, "The Great Book Massacre: An Exchange," *New York Review of Books* (March 14, 2002), www.nybooks.com/articles/15195.

65. See, for example, Vincent Kiernan, "'Open Archives' Project Promises Alternative to Costly Journals," *Chronicle of Higher Education*, December 3, 1999; Budapest Open Access Initiative, www.soros.org/openaccess. On questions of public domain and privatization, see Lawrence Lessig, *The Future of Ideas: The Fate of the Commons in a Connected World* (New York, 2001).

66. Justin Winsor, "Manuscript Sources of American History: The Conspicuous Collections Extant," *Papers of the American Historical Association* 3, no. 1 (1888): 9–27, www.historians.org/info/AHA_History/jwinsor.htm. On the central concern with teaching in schools, see Link, "American Historical Association, 1884–1984," 12–15.

2. Web of Lies? Historical Knowledge on the Internet

1. David Hochman, "In Searching We Trust," *New York Times*, March 14, 2004, section 9, 1.

2. Gertrude Himmelfarb, "A Neo-Luddite Reflects on the Internet," *Chronicle of Higher Education*, November 1, 1996, A56.

3. Gary J. Kornblith and Carol Lasser, eds., "Teaching the American History Survey at the Opening of the Twenty-First Century: A Round Table Discussion," *Journal of American History* 87 (March 2001), http://www.indiana.edu/~jah/textbooks/2001/ (accessed September 12, 2005).

4. Daniel J. Cohen, "By the Book: Assessing the Place of Textbooks in U.S. Survey Courses," *Journal of American History* 91 (March 2005): 1405–1415.

5. Paul Vitanyi and Rudi Cilibrasi, "Automatic Meaning Discovery Using Google," 2005, http://arxiv.org/abs/cs/0412098 (accessed April 30, 2005).

6. The most well-known of these programs in the National Institute of Standards and Technology's Text REtrieval Conference (TREC), also sponsored by the U.S. Department of Defense. See http://trec.nist.gov/.

7. Art Institute of Chicago, "Claude Monet's The Artist's House at Argenteuil, 1873," http://www.artic.edu/artexplorer/search.php?tab=2&resource=406 (accessed June 21, 2005).

8. Academie Des Beaux-Arts, "La Fondation Claude Monet à Giverny," http://www.academie-des-beaux-arts.fr/uk/fondations/giverny.htm (accessed June 21, 2005).

9. John A. Garraty and Eric Foner, eds., *The Reader's Companion to American History* (New York: Houghton Mifflin, 1991).

10. David W. Walker, "Who am I?" http://users.cs.cf.ac.uk/David.W.Walker/who.html (accessed September 12, 2005).

11. Forrest McDonald, "Hamilton, Alexander," in *American National Biography Online* (New York: Oxford University Press, 2000); New-York Historical Society, "Alexander Hamilton: The Man Who Made Modern America," http://www.alexander hamiltonexhibition.org/timeline/timeline1.html (accessed September 12, 2005).

12. James Surowiecki. *The Wisdom of Crowds: Why the Many Are Smarter Than the Few and How Collective Wisdom Shapes Business, Economies, Societies and Nations* (New York: Doubleday, 2004).

13. Jonathan Brent and Vladimir Naumov, *Stalin's Last Crime: The Plot Against the Jewish Doctors, 1948–1953* (New York: HarperCollins, 2003).

14. Much of this theory of information distance grows out of the work of A. N. Kolmogorov, "Three Approaches to the Quantitative Definition of Information," *Problems in Information Transmission* 1, no. 1 (1965): 1–7.

15. Our thanks to James Sparrow for pointing this out to us and for his other helpful comments on this article.

16. Charles C. Mann, *1491: New Revelations of the Americas Before Columbus* (New York: Knopf, 2005).

17. Justin Mullins, "Whatever Happened to Machines That Think?" *New Scientist* (April 23, 2005).

18. These are reported numbers, and subject to some debate. See Search Engine Watch, "Search Engine Sizes," http://searchenginewatch.com/reports/article.php/2156481 (accessed September 12, 2005); Caslon Analytics, "Net Metrics & Statistics Guide," http://www.caslon.com.au/metricsguide2.htm (accessed September 12, 2005).

19. Ben Elgin, "Revenge of the Nerds—Again," *BusinessWeek Online* (July 28, 2005), http://www.businessweek.com/technology/content/jul2005/tc20050728_5127_tc024.htm (accessed September 12, 2005).

20. Sam Wineburg, *Historical Thinking and Other Unnatural Acts: Charting the Future of Teaching the Past* (Philadelphia: Temple University Press, 2001), vii.

21. Sam Wineburg, "Crazy for History," *Journal of American History* 90 (March 2004): 1401–1414.

22. Open Content Alliance, *Open Content Alliance*, 2005, http://www.opencontentalliance.org/.

23. Some historical sociologists have pointed the way to doing such research using the coding of texts and content analysis. See Mady Wechsler Segal and Amanda Faith Hansen, "Value Rationales in Policy Debates on Women in the Military: A Content Analysis of Congressional Testimony, 1941–1985," *Social Science Quarterly* 73 (June 1992): 296–309; Paul Burstein, Marie R. Bricher, and Rachel L. Einwohner, "Policy Alternatives and Political Change: Work, Family, and Gender on the Congressional Agenda, 1945–1990," *American Sociological Review* 60 (February 1995): 67–83; Wendy Griswold, "American Character and the American Novel: An Expansion of Reflection Theory in the Sociology of Literature," *American Journal of Sociology* 86 (1981): 740–765; William Gamson and Andre Modigliani, "The Changing Culture of Affirmative Action," *Research in Political Sociology* 3 (1987): 137–77.

24. In a recent grant proposal to the National Endowment for the Humanities, Gregory Crane proposes to do just that: analyze and develop advanced linguistic and

statistical tools for the humanities ("An Evaluation of Language Technologies for the Humanities," draft of June 25, 2005).

25. William G. Thomas, III, "Computing and the Historical Imagination," in *A Companion to Digital Humanities*, ed. Susan Schreibman, Raymond George Siemens, and John Unsworth (Malden, Mass.: Blackwell, 2005), 56–68.26. In Thomas, "Computing and the Historical Imagination," 56.

27. A good introduction to these theories for humanists is Dominick Widdows, *Geometry and Meaning* (Stanford, Calif.: CLSI Publications, 2005).

3. *Wikipedia*: Can History Be Open Source?

1. My thanks to Melissa Beaver of the *Journal of American History* for compiling these figures. The 32,000 works include about 7,000 dissertations, which are never coauthored, but they also include coedited books, which involve a lower level of collaboration than coauthored books or articles.

2. See Richard Hofstadter, *The Age of Reform: From Bryan to F.D.R.* (New York, 1955), 131–73.

3. http://en.Wikipedia.org/wikistats/EN/TablesArticlesTotal.htm (Sept. 5, 2005). This count covers the period from the creation of the article on Franklin D. Roosevelt in September 2001 through July 4, 2005. See http://en.wikipedia.org/wiki/Franklin_Delano_Roosevelt. I am citing *Wikipedia* articles by URL and indicating the date accessed in parentheses because the articles continually change; readers can access the version I used by selecting the "history" tab and viewing the version from that date. All undated online resources were available when checked on Dec. 27, 2005.

4. Latest available numbers on visitors are for October 2004. The "official article count" for November 2005 is 2.9 million, 866,000 of them in English, according to http://en.wikipedia.org/wikistats/EN/TablesUsageVisits.htm (March 14, 2006). But the English-language home page says 1,023,303 articles. See http://en.wikipedia.org/wiki/Main_Page (March 14, 2006). Alexa rankings (available at http://www.alexa.com/) are from March 14, 2006. Information on number of employees was provided by Terry Foote (one of the employees) at a Hewlett Foundation meeting in Logan, Utah, on Sept. 27, 2005. See also Wikimedia Foundation, "Budget/2005," http://wikimediafoundation.org/wiki/Budget/2005 (Oct. 23, 2005). The statements of praise are quoted in Robert McHenry, "The Faith-Based Encyclopedia," TCS: Tech Central Station, Nov. 15, 2004, http://www. techcentralstation.com/111504A.html. For "joke," see Peter Jacso, "Peter's Picks and Pans," *Online* 26 (March 2002): 74.

5. There were c. 512 million words in May 2005, including 202 million in English. See http://en.wikipedia.org/wikistats/EN/TablesDatabaseWords.htm (Sept. 5, 2005).

6. http://en. wikipedia.org/wiki/History_of_Wikipedia (July 29, 2005); McHenry, "Faith-Based Encyclopedia"; http://www.gnu.org/encyclopedia/free-encyclopedia.

html. On Jimmy Wales, see Daniel Pink, "The Book Stops Here," *Wired* 13 (March 2005), http://www.wired.com/wired/archive/13.03/wiki_pr.html; Cynthia Barnett, "Wiki Mania," *Florida Trend* 48 (Sept. 2005), http://www.floridatrend.com/issue/default.asp?a=5617&s=1&d=9/1/2005; and Jonathan Sidener, "Everyone's Encyclopedia," SignOnSanDiego.com (Dec. 6, 2004) http://www.signonsandiego.com/union trib/20041206/news_mz1b6encyclo.html; http://en.wikipedia.org/wiki/Jimmy_Wales (July 5, 2005). On Larry Sanger, see Wade Roush, "Larry Sanger's Knowledge Free-for-All: Can One Balance Anarchy and Accuracy?," *Technology Review* 108 (Jan. 2005): 21; http://en.wikipedia.org/wiki/Larry_Sanger (Sept. 5, 2005). For their joint participation in Usenet groups, see Google Groups "humanities.philosophy.objectivism" and "alt.philosophy.objectivism." Until 2003, Bomis, in effect, owned *Wikipedia,* but in June of that year, all the assets were transferred to the nonprofit Wikimedia Foundation: http://en.wikipedia.org/wiki/Bomis (Oct. 29, 2005).

7. Larry Sanger, "The Early History of *Nupedia* and *Wikipedia:* A Memoir," *Slashdot,* April 18, 2005, http://features.slashdot.org/features/05/04/18/164213.shtml; http://en.wikipedia.org/wiki/History_of_Wikipedia (July 29, 2005).

8. http://en.wikipedia.org/wiki/Wikipedia:Multilingual_ranking_July_2005 (Aug. 16, 2005); http://en.wikipedia.org/wiki/History_of_Wikipedia (July 29, 2005).

9. http://en.wikipedia.org/wikistats/EN/TablesWikipediansContributors.htm (Sept. 1, 2005).

10. http://en.wikipedia.org/wiki/Wikipedia:Policies_and_guidelines (July 5, 2005).

11. Ibid.; http://en.wikipedia.org/wiki/No_original_research (July 5, 2005).

12. http://en.wikipedia.org/wiki/ NPOV (July 8, 2005).

13. Peter Novick, *That Noble Dream: The "Objectivity Question" and the American Historical Profession* (Cambridge, Eng., 1988), 3; http://en.wikipedia.org/wiki/Armenian_Genocide (July 10, 2005).

14. http://en.wikipedia.org/wiki/NPOV (July 8, 2005); http://en.wikipedia.org/wiki/Daniel_Pipes (Aug. 21, 2005).

15. http://en.wikipedia.org/wiki/Wikipedia:Policies_and_guidelines (July 5, 2005).

16. Free Software Foundation, *GNU Free Documentation License,* last modified May 2, 2005, http://www.fsf.org /licensing/licenses/fdl.html.

17. Daniel J. Cohen, "From Babel to Knowledge: Data Mining Large Digital Collections," *D-Lib Magazine* 12 (March 2006), http://www.dlib.org/dlib/march06/cohen/03cohen.html (March 21, 2006). For H-Bot, which was also developed by Daniel J. Cohen, see http://chnm.gmu.edu/tools/h-bot/ (March 21, 2006). For definitions of free software, see Free Software Foundation, "The Free Software Definition," http://www.fsf.org/licensing/essays/free-sw.html (March 21, 2006).

18. http://en.wikipedia.org/wiki/Wikipedia:Policies_and_guidelines (July 5, 2005).

19. Sanger, "Early History of *Nupedia* and *Wikipedia.*"

20. http://en.wikipedia.org/wiki/Wikipedia:Administrators (Sept. 5, 2005).

21. http://en.wikipedia.org/wiki/Wikipedia:Banning_policy (Sept. 5, 2005); http://en.wikipedia.org/ wiki/Jimmy_Wales (July 5, 2005). But note that Wales "has stated that if the two members of the board who edit *Wikipedia* vote the same way on something, he will cast his vote in their favor, effectively giving them the controlling majority." Ibid. Most of the money supporting the Wikimedia Foundation has come from successful fund-raising drives, but it has also received support from corporations and foundations. See Wikimedia Foundation, http://wikimediafoundation.org/wiki/Home. Wales also controls a for-profit company, Wikia, which sells ads; manages Wikicities, a collection of over 250 wiki communities; and hosts *Memory Alpha*, a Star Trek encyclopedia, and *Uncyclopedia*, a parody encyclopedia. http://en.wikipedia.org/wiki/Wikia (Dec. 28, 2005). See also Barnett, "Wiki Mania."

22. "Critical Views of *Wikipedia*," *Wikinfo* (a fork of *Wikipedia*), http://www.wikinfo.org/wiki.php?title=Critical_views_of_Wikipedia (July 23, 2005); http://en.wikipedia.org/wiki/Talk:Charles_Coughlin (Sept. 5, 2005). The article was locked when I first looked at it on Aug. 24, 2005, but it was unlocked on Sept. 1, 2005, with the comment that the "page has been protected for far too long. It's a wiki, time to let people edit it again." On edit wars, see Sarah Boxer, "Mudslinging Weasels into Online History," *New York Times*, Nov. 10, 2004, E1. But press accounts have tended to exaggerate the degree to which pages are locked. See http://en.wikipedia.org/wiki/Lyndon_LaRouche (Sept. 5, 2005); and the very extensive debate about the entry at http://en.wikipedia.org/wiki/Talk:Lyndon_LaRouche (Sept. 5, 2005).

23. http://en.wikipedia.org/wiki/History_of_the_United_States_(1918–1945) (July 31, 2005); Alan Brinkley, *Voices of Protest: Huey Long, Father Coughlin, and the Great Depression* (New York, 1983), 57–61, 108.

24. http://en.wikipedia.org/wiki/History_of_the_United_States_(1865–1918) (July 31, 2005); http://en.wikipedia.org/wiki/Feminist_history_in_the_United_States (July 31, 2005); http://en.wikipedia.org/wiki/Immigration_to_the_United_States (July 31, 2005). The sentence was lifted from the United States Information Agency's (USIA) online history textbook, available at http://odur.let.rug.nl/~usa/H/index.htm.

25. http://en.wikipedia.org/wiki/Cultural_history_of_the_United_States (July 31, 2005); http://en.wikipedia.org/wiki/Postage_stamps_and_postal_history_of_the_United_States (Sept. 18, 2005).

26. Taylor Branch, *Parting the Waters: America in the King Years, 1954–63* (New York, 1988); Taylor Branch, *Pillar of Fire: America in the King Years, 1963–65* (New York, 1998); Taylor Branch, *At Canaan's Edge: America in the King Years, 1965–68* (New York, 2006).

27. *American National Biography* received at least $2.5 million in grants. See Janny Scott, "Commerce and Scholarship Clash; Publisher Seeks to Update a Classic, to Cries of 'Thuggery,'" *New York Times*, Nov. 22, 1996, B1. For the travails of the encyclopedia business, see Ronna Abramson, "Look under 'M' for Mess," *Industry Stan-*

dard, April 9, 2001, 56; May Wong, "Pity the Poor Encyclopedia," *Associated Press*, March 6, 2004, accessed through Lexis-Nexis. On the development of *Encarta* (which involved a team of 135 people even in its first phase as a CD-ROM), see Fred Moody, *I Sing the Body Electric: A Year with Microsoft on the Multimedia Frontier* (New York, 1995), 6–17. Initially, *Encarta* was based on the mediocre Funk & Wagnall's encyclopedia, but massive revision has greatly improved it.

28. I averaged 9 biographies that were in all three sources (excluding that of Andrew Jackson, which distorted the comparison because of its unusual length): *American National Biography Online*: 1,552 words per biography; *Wikipedia*: 386; *Encarta*: 107. Of 20 Civil War army officers covered in *American National Biography Online*, *Wikipedia* had 8 and *Encarta* only 2.

29. http://en.wikipedia.org/wiki/Wikipedia:Why_Wikipedia_is_not_so_great (July 25, 2005); Jengold e-mail interview by Joan Fragaszy, June 4, 2004 (Center for History and New Media, George Mason University, Fairfax, Va.); Sanger, "Early History of *Nupedia* and *Wikipedia*."

30. Dale Hoiberg quoted in a good compilation of criticisms: http://en.wikipedia.org/wiki/Criticism_of_Wikipedia (Sept. 5, 2005).

31. http://en.wikipedia.org/wiki/Frederick_Law_Olmsted (April 10, 2005); Edward McManus, "Salomon, Haym," *American National Biography Online* (New York, 2000).

32. http://en.wikipedia.org/wiki/Frederick_Law_Olmsted (April 10, 2005); http://en.wikipedia.org/wiki/Andrew_Jackson_Downing (Aug. 16, 2005); http://en.wikipedia.org /wiki/Calvert_Vaux (Aug. 24, 2005). Professionally edited reference works also suffer from inconsistencies. The 1958 edition of the *Encyclopedia Britannica* repeats the Betsy Ross legend in the entry on her but not in the one on the flag, and has Pocahontas rescuing John Smith in the Smith entry but not the Pocahontas entry, according to Harvey Einbinder, *The Myth of the Britannica* (New York, 1964), 359–62, 179–80.

33. Michael J. McCarthy, "It's Not True about Caligula's Horse; Britannica Checked," *Wall Street Journal*, April 22, 1990, A1; http://en.wikipedia.org/wiki/Wikipedia:Errors_in_the_Encyclop%C3%A6dia_Britannica_that_have_been_corrected_in_Wikipedia (Sept. 5, 2005). The *American National Biography Online* entry (by William A. Nierenberg) gives the date of I. I. Rabi's Ph.D. as 1926, but *Encarta*, *Wikipedia*, *Dissertation Abstracts*, and the Columbia University catalog say 1927. Rabi submitted an article version of his dissertation to *Physical Review* in 1926, which may be the basis of *American National Biography Online*'s dating. See John S. Rigden, *Rabi: Scientist and Citizen* (Cambridge, Mass., 1987), 45.

34. Michael Kurzidim, "Wissenswettstreit. Die kostenlose Wikipedia tritt gegen die Marktführer Encarta und Brockhaus an" (Knowledge competition: Free *Wikipedia* goes head to head with market leaders *Encarta* and *Brockhaus*), *c't*, Oct. 4, 2004, 132–39; Jim Giles, "Internet Encyclopaedias Go Head to Head," *Nature*, Dec. 15, 2005,

http://www.nature.com/nature/journal/v438/n7070/full/438900a.html. The computer scientist Edward Felten compared 6 entries from *Wikipedia* to the similar articles in *Encyclopedia Britannica* and found that 4 of those in *Wikipedia* were better. Edward Felten, *Freedom to Tinker,* blog, Sept. 3, 2004, http://www.freedom-to-tinker.com/?p=674.

35. http://en.wikipedia.org/wiki/Abraham_Lincoln (Oct. 23, 2005); James McPherson, "Lincoln, Abraham," *American National Biography Online. Nature* comes to a similar conclusion about the writing in *Wikipedia,* noting that several of its expert readers found the articles "poorly structured and confusing." See Giles, "Internet Encyclopaedias Go Head to Head."

36. McManus, "Salomon, Haym." See also Alan Brinkley, "Roosevelt, Franklin D.," *American National Biography Online;* and T. H. Watkins, "Ickes, Harold," ibid.

37. http://en.wikipedia.org/wiki/William_Quantrill (July 15, 2005).

38. http://en.wikipedia.org/wiki/Calvin_Coolidge (March 10, 2006); http://en.wikipedia.org/wiki/Eugene_V._Debs (Aug. 24, 2005); *No Oil for Pacifists,* blog, June 17, 2005, http://nooilforpacifists.blogspot.com/2005/06/open-source-closed-minds.html.

39. http://en.wikipedia.org/wiki/Wikipedia:Why_Wikipedia_is_not_so_great (July 25, 2005); http://en.wikipedia.org/wiki/Talk:Frederick_Law_Olmsted (Sept. 5, 2005). The extreme case of this particular bias is the editing of *Wikipedia* entries to flatter oneself. The former MTV VJ (video jockey) Adam Curry anonymously edited the article on podcasting to emphasize his own contribution to it. Similarly, Jimmy Wales edited his own *Wikipedia* entry to remove references to Larry Sanger's role in cofounding the online encyclopedia and to Bomis Babes as presenting "pornography." *Wikipedia*'s guidelines on "autobiography" begin by quoting Wales: "It is a social faux pas to write about yourself." Daniel Terdiman, "Adam Curry Gets Podbusted," *Media Blog,* Dec. 2, 2005, http://news.com.com/2061-10802_3-5980758.html; http://www.cadenhead.org/workbench/news/2828 (Dec. 28, 2005); Rogers Cadenhead, "*Wikipedia* Founder Looks Out for Number 1," *Workbench,* Dec. 19, 2005, http://en.wikipedia.org/wiki/Talk:Jimmy_Wales (Dec. 28, 2005); http://en.wikipedia.org/wiki/Wikipedia:Autobiography#If_Wikipedia_already_has_an_article_about_you (Dec. 28, 2005). Wales later told a reporter that he regretted making the changes: "I wish I hadn't done it. It's in poor taste." Rhys Blakely, "Wikipedia Founder Edits Himself," *Times Online,* Dec. 20, 2005, http:// www.timesonline.co.uk/article/0,,1-1948005,00.html. In January 2006 it emerged that some congressional staffers were altering their bosses' biographies to provide more flattering portraits or, for example, to remove mentions of indicted House majority leader Tom DeLay. See http://en.wikinews.org/wiki/Congressional_staff_actions_prompt_Wikipedia_investigation (March 14, 2006).

40. http://en.wikipedia.org/wiki/Warren_G._Harding (July 4, 2005); http://en.wikipedia.org/wiki/-Franklin_Delano_Roosevelt (July 3, 2005). John Summers offers a concise case for the Warren G. Harding–Nan Britton affair in response to a let-

ter making the opposite case. Robert H. Ferrell and Warren G. Harding III to Editor, *Journal of American History* 88 (June 2001): 330–31; John Summers to Editor, ibid., 331–33.

41. Robert McHenry, "Whatever Happened to Encyclopedic Style?," *Chronicle of Higher Education,* Feb. 23, 2003, B13; Algernon Charles Swinburne quoted ibid.; Charles Van Doren quoted in Pink, "Book Stops Here."

42. William Emigh and Susan C. Herring, "Collaborative Authoring on the Web: A Genre Analysis of Online Encyclopedias," *Proceedings of the 38th Hawaii International Conference on System Sciences* (2005), http://csdl.computer.org/comp/proceedings/hicss/2005/2268 /04/22680099a.pdf.

43. http://en.wikipedia.org/wiki/Wikipedia:Replies_to_common_objections (July 23, 2005).

44. For a report that some larger-scale studies found similar rates of repairs of vandalism, see Pink, "Book Stops Here."

45. http://en.wikipedia.org/w/index.php?title=Syracuse,_New_York&diff=prev &oldid=5526247 (Dec. 27, 2005); Kathy Ischizuka, "The *Wikipedia* Wars: School Librarian Sparks Fight over Free Online Resource," *School Library Journal* 50 (Nov. 2004): 24; *Dispatches from the Frozen North,* blog, Sept. 4, 2004, http://www.frozen north.org/C2011481421/E652809545/; Simon London, "Web of Words Challenges Traditional Encyclopedias," *Financial Times,* July 28, 2004, 18. See also *The Now Economy,* blog, Sept. 8, 2004, http://blog.commerce.net/archives/2004/09/decentralized_a. html. A systematic study found that one common form of vandalism (mass deletion) is typically repaired within two minutes; see Fernanda B. Viegas, Marvin Wattenberg, and Kushal Dave, "Studying Cooperation and Conflict between Authors with History Flow Visualizations," IBM Watson Center Technical Report #04–19, 2004, http:// domino.research.ibm.com/cambridge/research.nsf/a1d792857da52f638525630f004e7ab 8/53240210b04ea0eb85256f7300567f7e?OpenDocument.

46. One administrator observed, "Basically the transaction costs for healing *Wikipedia* are less than those to harm it, over a reasonable period of time. I am an admin and if I see vandalism to an article, it takes about ten total clicks to check that editor has vandalized other articles and made no positive contributions, block the IP address or username, and rollback all of the vandalism by that user. It takes more clicks if they edited a lot of articles quickly, but they had to spend much more time coming up with stupid crap to put in the articles, hitting edit, submit, etc. After being blocked, they have to be really persistent to keep coming back to vandalize. Some are, but luckily many more people are there to notice them and revert the vandalism. Its a beautiful thing." Taxman 415a, comment on Sanger, "Early History of *Nupedia* and *Wikipedia,*" April 18, 2005, http://features.slashdot.org/comments.pl?sid=146479&threshold=1& commentsort=0&mode=thread&cid=12276095 (March 21, 2006). For number of edits per month, see http://en.wikipedia.org/wikistats/EN/TablesDatabaseEdits.htm (Sept. 5, 2005).

47. For *Wikipedia*'s own summary and links to some of the key coverage, see http:// en.wikipedia.org/wiki/John_Seigenthaler_Sr._Wikipedia_biography_controversy (Dec. 27, 2005). For the revised and corrected biography, see http://en.wikipedia.org/ wiki/John_Seigenthaler_Sr. (Dec. 27, 2005). For Seigenthaler's original op-ed article, see John Seigenthaler, "A False *Wikipedia* 'Biography,'" *USA Today,* Nov. 29, 2005, http://www.usatoday.com/news/opinion/editorials/2005-11-29-wikipedia-edit_x.htm.

48. Janet Kornblum, "It's Online, but Is It True?," *USA Today,* Dec. 6, 2005, http:// www.usatoday.com/tech/news/techpolicy/2005–12–06- wikipedia-truth_x.htm; Lessig quoted in Katharine Q. Seelye, "Rewriting History; Snared in the Web of a *Wikipedia* Liar," *New York Times,* Dec. 4, 2005, section 4, 1; Wade Roush, "*Wikipedia:* Teapot Tempest," *TR Blogs,* Dec. 7, 2005, http://www.technologyreview.com/Blogs/ wtr_15974,292,p1.html. Another recent change provides that some articles can be "semi-protected"—no one can make changes who has not been registered for at least four days. See http://en.wikipedia.org/wiki/Wikipedia:Semi-protection_policy (March 14, 2006).

49. See comments on *Dispatches from the Frozen North,* blog, Sept. 4, 2004.

50. McHenry, "Faith-Based Encyclopedia"; Seelye, "Rewriting History."

51. Pink, "Book Stops Here"; http://en.wikipedia.org/w/index.php?title= Watergate_scandal&action=history (Sept. 5, 2005). W. Mark Felt's name was inserted at 5:18 p.m. on May 31, 2005. Wynn Quon, "The New Know-It-All: Wikipedia Overturned the Knowledge Aggregation Model by Challenging Contributors to Constantly Improve Its Entries," *Financial Post,* Feb. 26, 2005, accessed through Lexis-Nexis. When the *Wall Street Journal* noted that *Wikipedia* was reporting an out-of-date figure for the number of Korean War dead, it was fixed the same day. Carl Bialik, "A Korean War Stat Lingers Long after It Was Corrected," *Wall Street Journal,* June 23, 2005, http://online.wsj.com/public/article/SB111937345541365397-C9Z_jEOnlmcAqp HtdvX4upR6r7A_20050723.html (March 14, 2006); http://en.wikipedia.org/w/ index.php?title=Korean_War&diff=15663363&oldid=15663246 (Sept. 4, 2005).

52. For the student's statement, see comments on Sanger, "Early History of *Nupedia* and *Wikipedia,*" April 18, 2005.

53. "Critical views of *Wikipedia,*" *Wikinfo* (a fork of *Wikipedia*), http://www. wikinfo.org/wiki.php?title=Critical_views_of_Wikipedia (July 23, 2005); John Morse comment on Clay Shirkey blog entry, "K5 Article on *Wikipedia* Anti-elitism," in *Many 2 Many,* Jan. 5, 2005, http://www.corante.com/many/archives/2005/01/03/k5_article_ on_Wikipedia_antielitism.php. See also John Morse, *Dystopia Box,* blog, http://dystopiabox.blogspot.com; http://www. answers.com/main/ir/about_company.jsp.

54. Liz Lawley, *Many 2 Many,* Jan. 4, 2005, http://www.corante.com/many/archives/2005/01/04/academia_and_wikipedia.php; Patrick McLean comment on Shirkey blog, Jan. 4, 2004, http://www.corante.com/many/archives/2005/01/03/k5_ article_on_Wikipedia_antielitism.php; Oedipa comment on Lawley, Jan. 5, 2005. See also Ischizuka, "*Wikipedia* Wars."

55. *American National Biography Online* does offer individual subscriptions for $89 per year; prices for institutions range from $495 to $14,000 per year depending on size. On open access to scholarship, see John Willinsky, *The Access Principle: The Case for Open Access to Research and Scholarship* (Cambridge, Mass., 2005).

56. http://en.wikipedia.org/wiki/Blocking_of_Wikipedia_in_mainland_China (Mar. 17, 2006). Although the Chinese have a highly sophisticated Internet filtering system, it is far from impermeable. An underground economy of proxy servers, a range of circumvention technologies, and anonymous Internet communication networks provide significant challenges to what has become known as the "great firewall of China."

57. http://en.wikipedia.org/wiki/Verifiability (July 26, 2005); http://en.wikipedia. org/wiki/Wikipedia:WikiProject_History (July 31, 2005).

58. http://en.wikipedia.org/wiki/Wikipedia:Abundance_and_redundancy (Aug. 30, 2005); http://en.wikipedia.org/wiki/Talk:Scopes_Trial (Sept. 6, 2005).

59. http://en.wikipedia.org/wiki/Wikipedia:What_is_a_featured_article (Sept. 1, 2005); http://en.wikipedia.org/wiki/Wikipedia:Peer_review (Sept. 1, 2005).

60. SJ e-mail interview by Fragaszy, June 22, 2005 (Center for History and New Media); Peter Myers, "Fact-Driven? Collegial? This Site Wants You," *New York Times,* Sept. 20, 2001, G2.

61. James W. Rosenzweig e-mail interview by Fragaszy, May 27, 2005 (Center for History and New Media); APWoolrich (*Wikipedia* username) e-mail interview by Fragaszy, May 27, 2005, ibid.

62. APWoolrich interview; Academic Challenger (*Wikipedia* username) e-mail interview by Fragaszy, May 6–26, 2005 (Center for History and New Media); Pink, "Book Stops Here."

63. I have, however, decided to refrain from editing *Wikipedia* entries until after I publish this article. Some of the hesitancy about participating in *Wikipedia* that professional historians have shown is captured in Richard Jensen's comment: "Ok, I confess, I write for *Wikipedia*." Although (as a reading of H-Net discussion lists makes clear) many professional historians remain skeptical about *Wikipedia,* there has been growing interest, and a growing number have begun to participate directly since mid-2005. Richard Jensen, "*Wikipedia* and the gape," online posting, h-shgape, Dec. 9, 2005, http://www.h-net.org/~shgape/.

64. Sanger, "Early History of *Nupedia* and *Wikipedia*"; http://en.wikipedia.org/wiki/User:JHK (Sept. 1, 2005). Some scientists who have edited *Wikipedia* entries on controversial subjects such as global warming have become involved in major battles. See Giles, "Internet Encyclopaedias Go Head to Head."

65. Wales quoted in http://en.wikipedia.org/wiki/No_original_research (July 5, 2005).

66. Ibid. The ban on "original research" is in some tension with the suggestion that "primary sources" be used in history articles. Presumably, Wikipedians would

3. *Wikipedia*: Can History Be Open Source? 265

support using primary sources to verify a particular fact but not to construct a new interpretation. Hence, you might use a primary source to verify that Franklin D. Roosevelt said "a date which will live in infamy," rather than "a date that will live in infamy," but not to decide whether he knew in advance about the attack on Pearl Harbor.

67. I borrow the idea of a "poetics" of history from Greg Dening, *History's Anthropology: The Death of William Gooch* (Lanham, Md., 1988), 2.

68. Roy Rosenzweig, "Marketing the Past: *American Heritage* and Popular History in the United States, 1954–1984," in *Presenting the Past*, ed. Susan Porter Benson, Stephen Brier, and Roy Rosenzweig (Philadelphia, 1986); http://en.wikipedia.org/wiki/Abraham_Lincoln (Oct. 23, 2005); McPherson, "Lincoln, Abraham."

69. http://en.wikipedia.org/wiki/Warren_G._Harding (July 4, 2005); http://en.wikipedia.org/wiki/Woodrow_Wilson (July 5, 2005).

70. http://en.wikipedia.org/wiki/The_Intimate_World_of_Abraham_Lincoln (Aug. 28, 2005); C. A. Tripp, *The Intimate World of Abraham Lincoln* (New York, 2005); http://en.wikipedia.org/wiki/Spanish-American_War (April 12, 2005); http://www.s-t.com/daily/02-98/02-15-98/a02wn012.htm; Kristin L. Hoganson, *Fighting for American Manhood: How Gender Politics Provoked the Spanish-American and Philippine-American Wars* (New Haven, 1998).

71. On "shared authority," see Michael Frisch, *A Shared Authority: Essays on the Craft and Meaning of Oral and Public History* (Albany, 1990).

72. Anonymous e-mail interview by Fragaszy, June 3, 2005 (Center for History and New Media).

73. Robert K. Merton, "The Normative Structure of Science," 1942, in *The Sociology of Science: Theoretical and Empirical Investigations,* by Robert K. Merton (Chicago, 1973), 275.

74. Yochai Benkler, "Coase's Penguin; or, Linux and the Nature of the Firm," *Yale Law Journal* 112 (Dec. 2002), http://www.benkler.org/CoasesPenguin.html.

75. *Clickworkers Results: Crater Marking Activity,* http://clickworkers.arc.nasa.gov/documents/crater-marking.pdf.

76. "Facts and Statistics," *FamilySearch Internet Genealogy Service,* July 1, 2003, http://www.familysearch.org/Eng/Home/News/frameset_news.asp?PAGE=home_facts.asp; "Free Internet Access to Invaluable Indexes of American and Canadian Heritage," *The Church of Jesus Christ of Latter-day Saints,* Oct. 23, 2002, http://www.lds.org/newsroom/showrelease/0,15503,3881-1-13102,00.html; http://www.familysearch.org; Benkler, "Coase's Penguin." On costs of digitizing, see Daniel J. Cohen and Roy Rosenzweig, *Digital History: A Guide to Gathering, Preserving, and Presenting the Past on the Web* (Philadelphia, 2005), 93, http://chnm.gmu.edu/digitalhistory/digitizing/4.php.

77. http://www.lexisnexis.com/academic/guides/southern_hist/plantations/plantj1.asp; Max J. Evans, "The Invisible Hand and the Accidental Archives," paper presented

at "Choices and Challenges Symposium," Henry Ford Museum, Oct. 8, 2004, http://www.thehenryford.org/research/publications/symposium2004/papers/evans.pdf (March 15, 2006).

78. More than 55,000 people have made at least ten edits. http://en.wikipedia.org/wikistats/EN/Tables-WikipediansContributors.htm (Sept. 11, 2005). Benkler, "Coase's Penguin."

79. Aaron Krowne, "The FUD-Based Encyclopedia," *Free Software Magazine*, March 2005 http://www.freesoftwaremagazine.com/free_issues/issue_02/fud_based_encyclopedia/. There is also danger that the entry owner may not be the person with the greatest expertise or best judgment. The PlanetMath FAQ acknowledges that "currently there is no real recourse for someone who, say, writes poor entries and refuses all corrections, out of spite. In the future this will be handled by a ratings system, and filtering/sorting based on rating." But it also notes that "we have had no problem along these lines as of yet." http://planetmath.org/?method=12h&from=collab&id=35&op=getobj (March 30, 2006). For more on Krowne's approach, see Aaron Krowne, "Building a Digital Library the Commons-Based Peer Production Way," *D-Lib Magazine* 9 (Oct. 2003), http://www.dlib.org/dlib/october03/krowne/10krowne.html (March 21, 2006).

5. Rewiring the History and Social Studies Classroom: Needs, Frameworks, Dangers, Proposals

Some of the material in this article was drawn from "Teaching Culture, Learning Culture, and New Media Technologies," by Randy Bass and Bret Eynon, an introductory essay for the volume, "Intranational Media: The Crossroads Conversations on Learning and Technology in the American Culture and History Classroom," *Works & Days* (Spring/Fall 1998). The authors are indebted to Bret Eynon, Deputy Directory of the American Social History Project/Center for Media and Learning, for his contributions both to the earlier text and, more significantly, to the ideas behind this article, which are in large ways the product of his collaborations.

1. Quotes from 1880 and 1881 *Scientific American* in Steven Lubar, *InfoCulture: The Smithsonian Book of Information Age Inventions* (Boston: Houghton Mifflin, 1993), 130, and Claude S. Fischer, *America Calling: A Social History of the Telephone to 1940* (Berkeley: University of California Press, 1992), 2 (see also 1 and 26). Carolyn Marvin, *When Old Technologies Were New: Thinking about Electrical Communication in the Late Nineteenth Century* (New York: Oxford University Press, 1988), and Graham Rayman, "Hello, Utopia Calling?" in *Word* (no date), http://www.word.com/machine/jacobs/phone/index.html.

2. Rossetto, quoted in David Hudson, *Rewired: A Brief and Opinionated Net History* (Indianapolis: Macmillan Technical Publishers, 1997), 7; Al-Jumhuriya in R. J. Lambrose, "The Abusable Past," *Radical History Review* 70 (1998): 184.

3. Robert Lenzner and Stephen S. Johnson, "Seeing Things as They Really Are," *Forbes* (March 10, 1997), http://www.forbes.com/forbes/97/0310/5905122n.htm; Birkerts in "The Electronic Hive: Two Views:" "Refuse It" (Sven Birkerts), and "Embrace It" (Kevin Kelly), *Harper's Magazine* (May 1994). See also Sven Birkerts, *The Gutenberg Elegies: The Fate of Reading in an Electronic Age* (Boston: Faber and Faber, 1994); and Todd Oppenheimer, "The Computer Delusion," *Atlantic Monthly* (July 1997).

4. Philip E. Agre, "Communities and Institutions: The Internet and the Structuring of Human Relationships," circulated through *Red Rock Eater's News Service,* copy available at http:egroups.com/group/rre/804.html.

5. Readers will note that many of our references here are to the teaching of history and American culture, since those are our own specialties, but we think that our arguments apply broadly to the curriculum area generally referred to as "social studies."

6. Sam Wineburg, "Making Historical Sense," in *Knowing, Teaching, and Learning History: National and International Perspectives*, ed. Peter Stearns, Sam Wineburg, and Peter Seixas (New York: NYU Press, 2000). On the question of factual knowledge, the most influential study of recent years has been Diane Ravitch and Chester Finn Jr., *What Do Our 17-Year-Olds Know? A Report on the First National Assessment of History and Literature* (New York: Harper & Row, 1987). There is a large literature debating the work of Ravitch and Finn. See, for example, William Ayers, "What Do 17-Year-Olds Know? A Critique of Recent Research," *Education Digest* 53 (Apr. 1988): 37–39; Dale Whittington, "What Have 17-Year-Olds Known in the Past?" *American Educational Research Journal* 28 (Winter 1991): 759–80; Deborah Meier and Florence Miller, "The Book of Lists," *Nation* 245 (Jan. 9, 1988): 25–27; Terry Teachout, "Why Johnny is Ignorant," *Commentary* (March 1988):69–71. There have been two more recent studies by the National Assessment of Education Progress (NAEP). For brief reports on these, see Michael Mehle, "History Basics Stump U.S. Kids, Study Finds," *Bergen Record,* April 3, 1990, A1; Carol Innerst, "History Test Results Aren't Encouraging; US Teens Flop on 'Basic' Quiz," *Washington Times,* November 2, 1995, A2.

7. Roy Rosenzweig and David Thelen, *The Presence of the Past: Popular Uses of History in American Life* (New York: Columbia University Press, 1998); see also http:chnm.gnm/edu/survey.

8. For Crossroads, see http://www.georgetown.edu/crossroads/; for New Media Classroom, which is co-sponsored by American Social History Project/Center for Media & Learning (ASHP/CML) in collaboration with the American Studies Association's Crossroads Project, see http://www.ashp/cuny.edu/index_new.html; for American Memory Fellows, see http://memory.loc.gov/ammem/ndlpedu/amfp/intro.html.

9. See http://www.nueva/pvt.k12.cn.us/~debbie/library/cur/20c/turn.html.

10. For WPA life histories, see http://lcweb2.loc.gov/ammem/ndlpedu/lesson97/first-hand/main.html; for use of free black registers, see http://historymatters.gmu.edu/text/3freeblacks-shulkin.html.

11. National Center for History in the Schools, *National Standards for United States*

History: Exploring the American Experience (Los Angeles: NCHS, 1994), 29. The new American Association of School Librarians standards for student learning similarly focus on information literacy, on the ability to find, select, analyze, and interpret primary sources. See "Information Power: The Nine Information Literacy Standards for Student Learning," http://www.ala.org/aasl/ip_nine.html.

12. The Oyez Project, Northwestern University, U.S. Supreme Court Multimedia Database, http://oyez.nwu.edu/; the U.S. Holocaust Memorial Museum, http://www.ushmm.org/; Liberty, Equality, Fraternity: Exploring the French Revolution is being developed by the Center for History and the New Media at GMU and the American Social History Project at CUNY and is available at http://chnm.gmu.edu/revolution.

13. Library of Congress, *Inventing Entertainment: The Early Motion Pictures and Sound Recordings of the Edison Companies,* http://memory.loc.gov/ammem/edhtml/ed home.html. For plans for National Gallery of Recorded Sound, see http://www.h-net.msu.edu/about/press/ngsw.html.

14. Ed Ayers, "The Futures of Digital History," unpublished paper delivered at the Organization of American Historians meeting, Toronto, April 1999.

15. On the "novice in the archive," see Randy Bass, "Engines of Inquiry: Teaching, Technology, and Learner-Centered Approaches to Culture and History," in American Studies Crossroads Project, *Engines of Inquiry: A Practical Guide for Using Technology in Teaching American Culture* (1997), which can be ordered from http://www.george town.edu/crossroads.

16. Sam Wineburg, "The Cognitive Representation of Historical 'Texts,'" in *Teaching and Learning in History,* ed. G. Leinhardt, I. L. Beck, and C. Stanton (Hillsdale, N.J., Erlbaum Associates, 1994), 85. See also Allan Collins, John Seeley Brown, and Ann Holum, "Cognitive Apprenticeship: Making Thinking Visible," *American Educator* (Winter 1991):6–11, 38–46.

17. At the collegiate level, one of the greatest advantages to using electronic interaction is that it increases the amount of time that students are focused on and interacting about the subject. Another advantage is the opportunity for asynchronous discussion: students can engage in the conversation on their own schedule, rather than only at the time when the instructor and other students are available. These uses are less pertinent at this time for the K–12 context than other benefits of online interaction that we describe.

18. See http://www.internet-catalyst.org/projects/PCG/postcard.html.

19. See http://www.fred.net/nhhs/html/newspage.html.

20. *Constructivism* implies a theory of learning that emphasizes the active creation of knowledge by the learner, rather than the imparting of information and knowledge by the instructor. A second meaning for constructivism, sometimes also called *constructionism,* is the extension of constructivist approaches that stresses the building of knowledge objects. "Constructionism," as defined by Yasmin Kafai and Mitchel Resnick, "suggests that learners are particularly likely to make new ideas when they are

actively engaged in making some type of external artifact . . . which they can reflect upon and share with others." Kafai and Resnick, eds., *Constructionism in Practice: Designing, Thinking, and Learning in a Digital World* (Mahwah, N.J.: Lawrence Erlbaum Associates, 1996), 1.

21.   http://www.fcps.k12.va.us/VirginiaRunES/museum/museum.htm;   Daniel Sipe, Presentation at NMC, New York, July 1997.

22. See http://www.wms-arl.org/amf1/student.htm.

23. See http://206.252.235.34/projects/local.htm. For a taxonomy of student constructive projects, with links to school and college-based examples, see http://www.georgetown.edu/crossroads/constructive.html.

24. On the problems of scientific forestry, see James C. Scott, *Seeing Like a State: How Certain Schemes to Improve the Human Condition Have Failed* (New Haven: Yale University Press, 1998), 11–22.

25. On history CD-ROMs, see Roy Rosenzweig, "So What's Next for Clio? CD-ROM and Historians," *Journal of American History* (March 1995):1621–1640.

26. Diane Ravich, "The Great Technology Mania," *Forbes* (March 23, 1998), http://www.forbes.com/forbes/98/0323/6106134a.htm.

27. U.S. Department of Education, National Center for Education Statistics, "Internet Access in Public Schools and Classrooms: 1994–98," (February 1999), http://nces.ed.gov/pubs99/1999017.html. See also National Telecommunications and Information Administration, U.S. Department of Commerce, *Falling Through the Net: Defining the Digital Divide* (July 1999), http://www.utia.doc.gov/ntiahome/digitaldivide/ and Paul Attewell and Juan Battle, "Home Computers and School Performance," *The Information Society* 15, no. 1 (1999): 1–10, which finds that students with computers at home have higher test scores even after controlling for family income but that children from high socioeconomic (and white) homes show larger educational gains with home computers than do lower SES (and minority) children.

28. For two examples of gateways see American Studies Electronic Crossroads (http://www.georgetown.edu/crossroads/) and History Matters: The U.S. Survey Course on the Web (http://historymatters.gmu.edu).

29. See Sam Wineburg, "Historical Thinking and Other Unnatural Acts," *Phi Delta Kappan* 80 (March 1999): 488–99.

7. Should Historical Scholarship Be Free?

1. National Institutes of Health, "Policy on Enhancing Public Access to Archived Publications Resulting from NIH-Funded Research," Feb. 3, 2005, http://grants.nih.gov/grants/guide/notice-files/NOT-OD-05–022.html. For detailed commentary, see SPARC *Open Access Newsletter* 82 (February 2, 2005), http://www.earlham.edu/~peters/fos/newsletter/02–02–05.htm.

2. *Budapest Open Access Initiative*, http://www.soros.org/openaccess.

3. See, for example, Stevan Harnad, "Scholarly Skywriting and the Prepublication Continuum of Scientific Inquiry," *Psychological Science* 1 (1990), http://www.ecs.soton. ac.uk/~harnad/Papers/Harnad/harnad90.skywriting.html; Suber's SPARC *Open Access Newsletter* and other links at http://www.earlham.edu/~peters/fos/; and various publications by Willinsky collected at http://pkp.ubc.ca/publications/index.html.

4. John Willinsky, "Copyright Contradictions in Scholarly Publishing," *First Monday* 7, no. 11 (November 2002), http://firstmonday.org/issues/issue7_11/Willinsky.

5. Richard Poynder, "No Gain Without Pain," *Information Today Online* 21, no. 10 (November 2004), http://www.infotoday.com/it/nov04/poynder.shtml.

6. http://www.historians.org/info/AHA_History/charter.htm.

7. Rudy Baum, "Socialized Science," *Chemical and Engineering News* 82, no. 38 (September 20, 2004), http://pubs.acs.org/cen/editor/8238edit.html. On general issue of scholarly societies, see John Willinsky, "Scholarly Associations and the Economic Viability of Open Access Publishing," *Journal of Digital Information* 4, no. 2 (April 9, 2003), http://jodi.ecs.soton.ac.uk/Articles/v04/i02/Willinsky/; Jim Pitman, "A Strategy for Open Access to Society Publications," January 28, 2004, http://stat-www.berkeley.edu/users/pitman/strategy.html; David C. Prosser, "Between a Rock and a Hard Place: The Big Squeeze for Small Publishers," *Learned Publishing* 17, no. 1 (2004), http://eprints.rclis.org/archive/00000945/01/Big_Squeeze.htm.

8. For a more systematic review, see John Willinsky, "The Nine Flavours of Open Access Scholarly Publishing," *Journal of Postgraduate Medicine* 49, no. 3 (2003), http://www.jpgmonline.com/article.asp?issn=0022–3859;year=2003;volume=49;issue=3;spage=263;epage=267;aulast=Willinsky.

9. Alma Swan, "Self-archiving: It's an Author Thing" (paper presented at Workshop on Open Access Institutional Repositories, University of Southampton, January 25, 2005), PowerPoint presentation available at www.eprints.org/jan2005/ppts/swan.ppt.

10. See http://www.biomedcentral.com/.

11. Poynder, "No Gain Without Pain."

12. John Willinsky, "Scholarly Associations."

8. Collecting History Online

Complete information for numbered links referred to in these notes may be found at http://chnm.gmu.edu/digitalhistory/links (accessed April 29, 2010).

1. MATRIX, H-Net: Humanities and Social Sciences Online, link 6.1.

2. Linda Shopes, "The Internet and Collecting the History of the Present" (paper presented at September 11 as History: Collecting Today for Tomorrow, Washington, D.C., September 10, 2003). For more on this "rapport" and the way rich historical

accounts arise during the live interaction of interviewer and interviewee, see Alessandro Portelli, *The Battle of Valle Giulia: Oral History and the Art of Dialogue* (Madison: University of Wisconsin Press, 1997) and Michael Frisch, *A Shared Authority: Essays on the Craft and Meaning of Oral and Public History* (Albany: State University of New York Press, 1991). It may also be worth comparing (or supplementing) the practical advice of this chapter with the offline advice of Donald A. Ritchie in *Doing Oral History* (Oxford: Oxford University Press, 2003), and Judith Moyer, "Step-by-Step Guide to Oral History," link 6.3.

3. *Moving Here: Two Hundred Years of Migration to England,* link 6.4a; BBC, *WW2 People's War,* link 6.4b.

4. National Park Foundation, "Rosie the Riveter Stories," Ford Motor Company Sponsored Programs, link 6.5a; *National Geographic,* Remembering Pearl Harbor, link 6.5b; Voices of Civil Rights, link 6.5c; Alfred P. Sloan Foundation, "History of Science and Technology," link 6.5d; C250 Perspectives: Write Columbia's History, link 6.5e; The Vietnam Project: The Oral History Project—How to Participate, link 6.5f.

5. Spencer Weart, "Icedrilling: History of Greenland Ice Drilling," Discovery of Global Warming, link 6.6.

6. Computer History Museum, Apple Computer History Weblog, link 6.7a; Andy Hertzfeld, Folklore.org: Macintosh Stories, link 6.7b; David Kirsch, Electronic Vehicle History Online Archive, link 6.7c; CHNM, Echo: Exploring and Collecting History Online—Science, Technology, and Industry, link 6.7d.

7. Kevin Roe, Brainerd, Kansas: Time, Place and Memory on the Prairie Plains, link 6.8a; Rowville Lysterfield History Project, Rowville Lysterfield History Project, link 6.8b; Exploratorium, Remembering Nagasaki: Atomic Memories, link 6.8c; SeniorNet, World War II Living Memorial, link 6.8d; "Veterans' Forums," History Channel, link 6.8e.

8. Joshua Greenberg, Video Store Project, link 6.9. The resulting dissertation was entitled "From Betamax to Blockbuster: Medium and Message in the Video Consumption Junction" (PhD. diss., Cornell University, 2004).

9. National Park Foundation, "Rosie the Riveter Stories"; BBC, "Associate Centres, WW2 People's War, link 6.10.

10. Keith Whittle, Atomic Veterans History Project, link 6.11.

11. Sixties Project, The Sixties-L Discussion List, link 6.12.

12. "Tell Your Story," Moving Here: Two Hundred Years of Migration to England, link 6.15.

13. See link 6.16.

14. Several companies offer inexpensive voicemail services that allow contributors to contact the collecting institution via a toll-free telephone number and record a message according to prompts you provide. These services can be configured so that recordings are e-mailed to you as digital audio attachments, complete with date, time, and incoming number stamp. From your e-mail inbox, these recordings can be easily

archived, made available on your Web site, or edited for other kinds of public presentation. UReach, Onebox, MaxEmail, and many other telecommunications companies provide such services for less than 10 cents per minute. Burgeoning technologies for making phone calls via the Internet, including so-called VoIP (Voice over Internet Protocol) services from most of the large telecommunications companies as well as start-ups such as Vosage and Skype, also hold promise for conducting and recording historical interviews cheaply, since the sound from these calls is already digital and can be stored on your hard drive using special software.

15. Jewish Museum, "The Jewish Lads' Brigade," Moving Here: Two Hundred Years of Migration to England, link 6.18.

16. CHNM, "Claude Shannon: The Man and His Impact," Echo: Exploring and Collecting History Online—Science, Technology, and Industry, link 6.19a. For the Siberia entry, see the submission dated August 5, 2001, link 6.19b.

17. Kirsch, Electric Vehicle History Online Archive.

18. National Institutes of Health, A Thin Blue Line: The History of the Pregnancy Test Kit, link 6.21.

19. James Sparrow, Blackout History Project, link 6.22.

20. A couple of years ago it seemed as if oral history had successfully won an exemption from the stricter IRB rules. See Bruce Craig, "Oral History Excluded from IRB Review," Perspectives (December 2003), link 6.23a; American Historical Association, "Questions Regarding the Policy Statement on Institutional Review Boards," Press Release, November 10, 2003, link 6.23b; Donald A. Ritchie and Linda Shopes, "Oral History Excluded from IRB Review," Oral History Association, link 6.23c. More recent developments have put this exemption in question. See Robert B. Townsend and Mériam Belli, "Oral History and IRBs: Caution Urged as Rule Interpretations Vary Widely," Perspectives (December 2004), link 6.23d. For general guidelines for ethically conducting interviews (online or off), see Oral History Association, Oral History Evaluation Guidelines, Pamphlet Number 3, September 2000, link 6.23e; American Historical Association, "Statement on Standards of Professional Conduct," May 2003, link 6.23f.

21. For good examples of short, clear terms of contribution and use, see the submission page for the Voice of Civil Rights project at link 6.24a or the Echo project policies page at link 6.24b; for further guidance on building a policies page, see "TRUSTe Model Privacy Disclosures," TRUSTe: Make Privacy Your Choice, link 6.24c.

22. "Tell Your Story,:" link 6.25a; National Park Foundation, "Rosie the Riveter Stories—Your Contact Information," Ford Motor Company Sponsored Programs, link 6.25b; National Park Foundation, "Rosie the Riveter Stories—'Your Stories' Terms of Submission and Disclaimer," Ford Motor Company Sponsored Programs, link 6.25c.

23. See R. Tourangeau, L. J. Rips, and K. Rasinski, *The Psychology of Survey Response* (New York: Cambridge University Press, 2000) for an overview of the subject. For more on Web surveys from the social science perspective, see M. P. Couper, M. Traugott, and M. Lamias, "Web Survey Design and Administration," *Public Opinion Quarterly* 65, no. 2 (2001): 230–53, and M. P. Couper, "Web Surveys: A Review of Issues and Approaches," *Public Opinion Quarterly* 64, no. 4 (2000): 464–94. A full bibliography of survey design is available from the Laboratory for Automation Psychology and Decision Processing at the Human/Computer Interaction Laboratory at the University of Maryland, link 6.26.

24. Quotation from Don A. Dillman, "Internet Surveys: Back to the Future," *The Evaluation Exchange* 10, no. 3 (2004): 6. See also Don A. Dillman, *Mail and Internet Surveys: The Tailored Design Method* (New York: Wiley, 2000), and related papers at link 6.27.

25. "Web Newspaper Registration Stirs Debate," CNN.com, 14 June 2004 [[[June 14, 2004?]]], link 6.28a. Online collecting projects that focus on sensitive topics obviously may encounter more resistance to revealing accurate personal information. See R. Coomber, "Using the Internet for Survey Research," *Sociological Research Online* 2, no. 2 (1997), link 6.28b.

26. The American Registry for Internet Numbers has a free IP lookup service at link 6.29a. Non-U.S. domains (those with two-letter country codes at the end) can be located through Uwhois.com, link 6.29b. Domains that end in .aero, .arpa, .biz, .com, .coop, .edu, .info, .int, .museum, .net, and .org can be located through the governing body for the Web, the Internet Corporation for Assigned Names and Numbers (ICANN), link 6.29c. Several commercial services scan worldwide IP addresses, e.g., Network-tools.com, link 6.29d, and Network Solutions, link 6.29e.

27. Michael Kazin, "12/12 and 9/11: Tales of Power and Tales of Experience in Contemporary History," *History News Network,* September 11, 2003, link 6.30.

28. Pew Internet and American Life Project, "One Year Later: September 11 and the Internet" (Washington, D.C.: Pew Internet and American Life Project, 2002). See also Bruce A. Williams and Michael X. Delli Carpini, "Heeeeeeeeeeeere's Democracy!" *Chronicle of Higher Education,* April 19, 2002, 14.

29. Lane Collins, Geoffrey Hicks, and Marie Pelkey, Where Were You: September 11th, 2001, link 6.32.

30. Both the September 11 project and the 2000 election project were launched under the auspices of the library's larger Web preservation effort named MINERVA (Mapping the INternet Electronic Resources Virtual Archive). See link 6.33a. For the projects, see Library of Congress, The September 11 Web Archive, link 6.33b; Library of Congress, Election 2002 Web Archive, link 6.33c. For an overview of collection statistics, see Library of Congress, "Welcome," The September 11 Web Archive, link 6.33d.

31. CHNM and the American Social History Project, The September Digital Archive, link 6.34.

32. Here Is New York: A Democracy of Photographs, link 6.35.

33. Pew Internet and American Life Project, *The Commons of the Tragedy* and *How Americans Used the Internet After the Terror Attack* (Washington, D.C.: Pew Internet and American Life Project, 2001); quotation from *The Commons of the Tragedy*. See also Amy Harmon, "The Toll: Real Solace in a Virtual World: Memorials Take Root on the Web," *New York Times,* September 11, 2002, G39. For more on the growth of Internet usage, especially as a place for communication, expression, and dialogue, see Deborah Fallows, *The Internet and Daily Life* (Washington, D.C.: Pew Internet and American Life Project, 2004).

34. Herodotus, *The History*, trans. David Grene (Chicago: University of Chicago Press, 1987), 35.

9. Brave New World or Blind Alley? American History on the World Wide Web

1. *Washington Post,* Jan. 14, 1996, H1; *Fort Lauderdale Sun-Sentinel,* Dec. 8, 1995, D1. On the origins of the Internet, see Katie Hafner and Matthew Lyon, *Where Wizards Stay Up Late: The Origins of the Internet* (New York, 1996); and Peter H. Salus, *Casting the Net: From Arpanet to Internet and Beyond . . .* (Reading, Mass., 1995).

2. Matthew Gray, "Web Growth Summary," http://www.mit.edu/people/mkgray/net/web-growth-summary.html; *New York Times,* Nov. 20, 1995, 1.

3. *New York Times,* Oct. 28, 1996, D10.

4. Gertrude Himmelfarb, "A Neo-Luddite Reflects on the Internet," *Chronicle of Higher Education,* Nov. 1, 1996, A56.

5. Randy Bass, "The Garden in the Machine: The Impact of American Studies on New Technologies," http://www.georgetown.edu/bassr/garden/html; Michael Joyce, "The Lingering Errantness of Place (In Memory of Sherman Paul)," paper delivered at the Association of College and Research Libraries/Library and Information Technology Association Joint Presidents Program at the annual conference of the American Library Association, Chicago, June 26, 1995, http://iberia.vassar.edu/~mijoyce/lingering_errantness.html.

6. Herbert I. Schiller, *Information Inequality: The Deepening Social Crisis in America* (New York, 1996), B6, 75.

7. Important history sites not discussed here include those organized around teaching courses. For example, see the New Media Classroom, http://www.gmu.edu/chnm/nmc; and the American Studies Crossroads Project, http://www.georgetown.edu/crossroads/webcourses.html. Andrew McMichael, Michael O'Malley, and Roy Rosenzweig, "Historians and the Web: A Guide," *Perspectives* 34 (Jan. 1996): 11–15, http://www.gmu.edu/chnm/beginner.html.

8. For a review of the comparative merits of the different search engines (which rates Excite as the best "general-purpose Web search site"), see Amarendra Singh and David Lidsky, "All-Out Search," *PC Magazine,* Dec. 3, 1996, 213 ff. For a discussion of the basic principles of search engines, see Steve G. Steinberg, "Seek and Ye Shall Find (Maybe)," *Wired* 4 (May 1996): 108 ff.

9. See http://www.yahoo.com/Arts/Humanities/History/U_S_History. To review the Web is to comment on a moving target. Statements about numbers of sites were true for the fall of 1996 when we wrote this article; the specific numbers surely have changed by the time you are reading this.

10. Michael A. Hoffman II, "The Campaign for Radical Truth in History," http://www.hoffman-info.com/; National Women's History Project, http://www.nwhp.org/; John Stanley, "One-Third of a Nation: Overview of a Living Newspaper," http://mason.gmu.edu/~jstanle1/FTP/index.html; Life History Manuscripts from the Folklore Project, Works Progress Administration (WPA) Federal Writers Project, 1936–1940, http://rs6.loc.gov/wpaintro/wpahome.html. In February 1997 the John Stanley link was no longer active.

11. http://www.webcom.com/teddy/topic.html; http://www.georgetown.edu/crossroads; http://www.gmu.edu/chnm; Ron San Juan, "Welcome to the *Miss Saigon* Page," http://www.clark.net/pub/rsjdfg/; Welcome to Magellan!, http://www.mckinley.com/.

12. AltaVista Technology Inc., http://www.altavista.com.

13. http://sunsite.berkeley.edu/Goldman/. The enterprise of searching can take on a slightly frightening "big brother" quality; a search for our own names comes up with references to our work not only in online syllabi but also in long-forgotten postings to an online bulletin board. In effect, every casual utterance on the Web is instantly archived and indexed for posterity.

14. Grolier Online: The American Presidency, http://gepweb1.grolier.com/presidents/ea/side/debs.html; Hoosiers—Individuals with Significant Ties to Indiana, http://doc.state.in.us/LearningResources/persons; Eugene V. Debs in West Virginia, 1913, http://www.wvlc.wvnet.edu/history/journal_wvh/wvh52.html; Debs Collection, http://odin.indstate.edu/level1.dir/rare.html#Debs.

15. The Web-based *Encyclopedia Americana* article, for example, was written more than thirty years ago by Ray Ginger (who died in 1975), although the copyright is given as 1996; see Grolier Online. Nick Salvatore, *Eugene V. Debs: Citizen and Socialist* (Urbana, 1982).

16. Carnegie, Andrew, http://www.lib.utexas.edu/Libs/HRC/WATCH/; Bridging the Urban Landscape: Andrew Carnegie, a Tribute, http://www.clpgh.org/exhibit/carnegie.html; Carnegie Club: The Carnegie Story, http://www.expressmedia.co.uk/carnegie/story.htm.

17. For the 110 million figure, see Linton Weeks, "Brave New Library," *Washington Post Magazine,* May 26, 1991, 11 ff.

18. For general background on the National Digital Library Program (NDLP), see

Caroline R. Arms, "Historical Collections for the National Digital Library: Lessons and Challenges at the Library of Congress," *D-Lib Magazine* (April, May 1996), http://lcweb2.loc.gov/ammem/dlib-2part.html.

19. "Periodic Report from the National Digital Library Program, the Library of Congress" (June 1996), 1. The quest to get collections digitized as inexpensively as possible raises thorny ethical and political problems. The Library of Congress has already faced criticism for subcontracting the digitization to prison labor and to electronic sweatshops in the Philippines and Jamaica. See Marcia Gelbart, "Hill Library Turns Back on 'Buy America,'" *Hill,* Sept. 11, 1996, 1.

20. Search times will vary according to the time of day. In addition, one must contend with technological failure; the second time we tried this search we got the message "This collection is temporarily unavailable. Please try later." The Central Park references include some excellent, but obscure, photos of the park, and three WPA life history interviews that talk about the park as a space for sexual encounters—a topic difficult to research in other sources.

21. Here are a few lines, as displayed, from one WPA life history: "{Begin deleted text} to {End deleted text} {Begin inserted text} {Begin handwritten} to {End handwritten} {End inserted text} get back {Begin deleted text} but {End deleted text} {Begin inserted text} {Begin handwritten} and {End handwritten} {End inserted text} I'd go under tryin. It'd be like life thatway. {Begin deleted text} you {End deleted text} {Begin inserted text} ya {End inserted text} wanna live but ya gotta die . . . {Begin deleted text} [?][?][?][?] {End deleted text} {Begin deleted text} [?][?][?] {End deleted text}." The alternative to attempting to read this is to download an actual image of the original document. This takes a few minutes but reads more easily. It is a boon to have both alternatives available. The lower quality of the images is due to earlier, automated digital transfers from film, which are presumably too expensive to redo.

22. In a single day of remote research in the National Digital Library, we had probably a dozen different technical problems—crashes in trying to use "RealAudio" from the Nation's Forum or cryptic messages like "Temporary file open error. Display failed" or "inquiry failed." But regular users of the Library of Congress know other sorts of technical problems well (slips marked "not on shelf" when the book is there, for example), and the Web, after all, offers a library that never closes and from which the books are never checked out to a congressman.

23. Making of America Project, http://library.cit.cornell.edu/MOA/moa-mission.html.

24. Gilder Lehrman Institute of American History, http://vi.uh.edu/pages/mintz/GILDER.HTM; Confederate Broadside Poetry Collection, http://www.wfu.edu/Library/rarebook/broads.htm; Library of Congress-Ameritech NDLP Competition, http://lcweb2.loc.gov/ammem/award/index.html.

25. Anti-Imperialism in the United States, 1898–1935, http://web.syt.edu:80/~fjzwick/ail98–35.html.

26. Valley of the Shadow: Two Communities in the American Civil War, http://jefferson.village.virginia.edu/vshadow2/.

27. Vannevar Bush, "As We May Think," *Atlantic Monthly* 76 (July 1945), http://www.ausbcomp.com/~bbott/wik/vbush.htm. Computer Museum Network, http://www.tcm.org/.

28. On the Electronic Numerical Integrator and Computer (ENIAC) and J. Presper Eckert, see Birth of the Information Age, http://homepage.seas.upenn.edu/~museum/; and Presper Eckert Interview, http://www.si.edu/resource/tours/comphist/eckert.htm.

29. Metropolitan Lives: The Ashcan Artists and Their New York, http://www.nmaa.si.edu/metlives/ashcan.html; Museum of the City of New York, http://www.netresource.com/mcny/home.html; Museum of the City of San Francisco, http://www.sfmuseum.org/. The University of California's Museum of Paleontology's useful "Subway system" guide to the Web lists more than eighty history museums and forty libraries with Web sites and online exhibits. Art, History, Culture, and Science Museums, http://www.ucmp.berkeley.edu/subway/art.html.

30. Chetro Ketl Great Kiva 3-D Model Home Page, http://www.sscf.ucsb.edu/anth/projects/great.kiva/index.html.

31. HistoryNet, http://www.TheHistoryNet.com.

32. There is one page of links to "History On-Line," but it is a somewhat limited and eclectic list of history Web sites rather than a comprehensive guide.

33. Discovery Channel, http:/discovery.com; Rajiv Chandrasekaran, "A Top-Dollar Web Service Awaits Rerun," *Washington Post Business,* Nov. 4, 1996, 19. The story of Anne Bonny and Mary Read was written by Shay McNeal, an independent historical writer and publisher with her own Web site on colonial America, Wild Women and Salty Dogs, http://www.discovery.com/doc/1012/world/history/2pirates/2pirates.html. McNeal and Poullin Historical Publishing, http://www.colonialhist.inter.net/Welcome.html; H. J. Fortunato, "From Here to Obscurity," http://discovery.com/doc/1012/world/obscurity/obscurity042296/obscurity.html.

34. Lori Ann Wark, "The Day of the Black Blizzard," http://www.discovery.com/DCO/doc/1012/world/history/dustbowl/dustbowlopener.html; History 409: The Dust Bowl, http://chnm.gmu.edu/hist409/dust.html.

35. For the term "infotainment" see Benjamin R. Barber, *Jihad vs. McWorld* (New York, 1995). History Channel, http://www.historychannel.com.

36. American Civil War Homepage, http://funnelweb.utcc.utk.edu/~hoemann/cwarhp.html; Larry Stevens to Roy Rosenzweig, e-mail, Nov. 6, 1996. Another large category of historical Web sites includes those created by Web design firms and Internet service providers to advertise their services. See, for example, the North Georgia

History site created by Golden Ink Web Design Services and the Negro Leagues Baseball On-Line Archives, which was designed and maintained by MoxieWeb, a Web design and hosting firm. http://www.ngeorgia.com/history/; http://www.infi.net/~moxie/nlb/nlb.html.

37. Winsor McCay Page!, http://pandorasbox.com/littlenemo.html; Time Exposure: The On-Line Bibliography of Web-based Information About William Henry Jackson, http://www.fit.edu/InfoTechSys/resources/cogsei/white.html; Money–Past, Present, & Future, http://www.ex.ac.uk~RDavis/arian/money.html; Houdini!, http://www.uelectric.com/houdini/contents.htm.

38. John Yu to Roy Rosenzweig, e-mail, Oct. 25, 1996; Japanese American Internment, http://www.geocities.com/Athens/8420/main.html; Unofficial Nikkei Home Page, http://www.kent.wednet.edu/KSD/SJ/Nikkei/Nikkei_homePage.html: Japanese Internment, Santa Clara Valley: On-Line Exhibit, http://scuish.scu.edu/SCU/Programs/Diversity/exhibit1.html; Manzanar Project by: Mark Leck and Doug Lockert, www.mvhs.srvusd..k12.ca.us/~mleck/man/; War Relocation Authority Camps in Arizona, 1942–1946, http://dizzy.library.arizona.edu/images/jpamer/wrainintro.html; Poston, Arizona, 1942–1996, http://www.u.arizona.edu/~scooter/home/post/post.html.

39. *New York Times,* Dec. 10, 1996, D4.

10. Wizards, Bureaucrats, Warriors, and Hackers: Writing the History of the Internet

1. I checked the indexes of the following seven books for references to "ARPA," "ARPANET," "computer," "IBM," or "Internet," and only found references to computers (but not the Internet) in the Schaller volume: William H. Chafe, *The Unfinished Journey: America Since World War II* (New York, 1995); Otis L. Graham Jr., *A Limited Bounty: The United States Since World War II* (New York, 1996); George Donelson Moss, *Moving On: The American People Since 1945* (Englewood Cliffs, 1994); Frederick F. Siegel, *Troubled Journey: From Pearl Harbor to Ronald Reagan* (New York, 1984); Joseph Siracusa, *The Changing of America: 1945 to the Present* (Arlington Heights, Ill., 1986); Michael Schaller, Virginia Scharff, Robert Schulzinger, *Present Tense: The United States Since 1945* (Boston, 1992); Howard Zinn, *Postwar America: 1945–1971* (Indianapolis, 1973). For pre–1988 coverage, see David Burnham, "Reagan Seeks Drive to Raise Productivity of U.S. Agencies," *New York Times,* February 20, 1985, A18. The Internet got its first real notice in the mainstream media in November 1988 when Robert Morris's "virus" temporarily shut it down: John Markoff, "Author of Computer 'Virus' Is Son Of N.S.A. Expert on Data Security," *New York Times,* November 5, 1988, A1. For a perceptive counter to the utopian language that often surrounds discussions

of the Internet, see Phil Philip E. Agre, "Yesterday's Tomorrow" (1998), http://dlis. gseis.ucla.edu/people/pagre/tls.html (a slightly different version was also published in the *Times Literary Supplement*, July 3, 1998).

2. For reviews of the historiography, see, for example, John M. Staudenmaier, *Technology's Storytellers: Reweaving the Human Fabric* (Cambridge, Mass., 1985), which argues that at least half the articles in *Technology and Culture*'s first two decades of publication take a "contextual" approach, and Stephen H. Cutliffe and Robert C. Post, eds., *In Context: History and the History of Technology: Essays in Honor of Melvin Kranzberg* (Bethlehem, Pa., 1989). For a perceptive overview of writing in computer history, see Michael S. Mahoney, "The History of Computing in the History of Technology," *Annals of the History of Computing* 10, no. 2 (1988): 113–25.

3. Katie Hafner and Matthew Lyon, *Where Wizards Stay Up Late: The Origins of the Internet* (New York, 1996).

4. BBN did not, however, exercise any control over the actual book. I have used the abbreviation ARPA throughout this essay, but, in fact, it later became the Defense Advanced Research Projects Agency (DARPA) and in 1993, it became ARPA again. A key initial focus of ARPA was space exploration, but that work was soon spun off into NASA.

5. Hafner and Lyon, *Where*, 12–13, 42.

6. Hafner and Lyon, *Where*, 44, 25, 74, 92, 102; Peter H. Salus, *Casting the Net: From ARPANET to Internet and Beyond* (Reading, Mass., 1995), 34.

7. Bruce Sterling "A Brief History of the Internet," *The Magazine of Fantasy and Science Fiction* (February 1993), but found online at http://www.forthnet.gr/forthnet/ isoc/short.history.of.internet. This account is also conventionally given (albeit sometimes in garbled form) in the many technical manuals on the Internet. See, for example, *The Internet Unleashed, 1996* (Indianapolis, 1995), 10, which begins its history of the Net with the heading: "From the Cold War—A Hot Network."

8. On Davies's work, see Martin Campbell-Kelly, "Data Communications at the National Physical Laboratory (1965–1975)," *Annals of the History of Computing* 9 (1988): 221–47.

9. Hafner and Lyon, *Where*, 56. On Rand and Herman Kahn, see Fred Kaplan, *The Wizards of Armageddon* (New York, 1983), 220–31.

10. Quoted in Arthur L. Norberg and Judy O'Neill with contributions by Kerry J. Freedman, *Transforming Computer Technology: Information Processing for the Pentagon, 1962–1986* (Baltimore, 1996), 166. According to Taylor, he was initially unaware of Baran's work, but Janet Abbate points out that "Baran's ideas quickly entered networking discourse and practice" and that Baran "discussed his ideas with many computing and communications experts and his report was widely read by others." Janet Abbate, "From Arpanet to Internet: A History of Arpa-Sponsored Computer Networks, 1966–1988," Ph.D. thesis, Univ. of Penn., 1994, 27.

11. Hafner and Lyon, *Where*, 79–80.

12. Martin Campbell-Kelly and William Aspray offer a very good, but brief, version of this analysis in *Computer: A History of the Information Machine* (New York, 1996), 283–94.

13. Hafner and Lyon, *Where*, 176.

14. Norberg and O'Neill, *Transforming*, vii. In 1986 IPTO was restructured and became the Information Science and Technology Office.

15. Norberg and O'Neill, *Transforming*, 6, 14, 25, 66.

16. The office was, in fact, initially called the Command and Control Division.

17. Norberg and O'Neill, *Transforming*, 12, 29. Still, there is a difficult problem here of sorting out rhetoric from reality. Abbate maintains that "the agency's disavowal of basic research was more rhetorical than real" and that while "resulting technologies often became part of the military command and control system, the defense rationale may have come after the fact." "From Arpanet," 77.

18. Norberg and O'Neill, *Transforming*, 163, 193. They also trace back the networking experiment to Licklider's desire to foster "community" among the researchers funded by ARPA (154). This point is particularly stressed in Judy O'Neill, "The Role of ARPA in the Development of the ARPANET, 1961–1972," *Annals in the History of Computing* 17 (1995): 76–81.

19. In 1969, for example, Congress passed a rider—the Mansfield Amendment—to the military reauthorization bill that mandated that "None of the funds authorized to be appropriated by this Act may be used to carry out any research project or study unless such project or study has a direct or apparent relationship to a specific military function or operations." Norberg and O'Neill, *Transforming*, 36.

20. For Robert Kahn's relationship to Herman, see "An Interview with Robert E. Kahn," conducted by Judy O'Neill, April 24, 1990, Reston, Virginia, Charles Babbage Institute, Center for the History of Information Processing, University of Minnesota, Minneapolis.

21. Hafner and Lyon, *Where*, 223.

22. Hafner and Lyon, *Where*, 251, 258.

23. Salus, *Casting the Net*, 126.

24. Norberg and O'Neill, *Transforming*, 20.

25. Paul N. Edwards, *The Closed World: Computers and the Politics of Discourse in Cold War America* (Cambridge, Mass., 1996), xv. John Staudenmaier notes the importance for historians of technology of a "master narrative" that offers a "whig reading of Western technological evolution as inevitable and autonomous." He also observes a generational divide in which younger scholars have "argued for a reading of the sometimes technically irrational dimensions of technological decision making as politically or culturally motivated and of the concept of progress in particular as a conceptual tool that helps technical elites to dominate their inferiors." Although Edwards's work is more influenced by Foucault and cultural studies than by the history of technology, his book clearly fits with those emphasizing the "dark side" of technology. Staudenmaier, "Recent Trends in the History of Technology," *American Historical Review* 95

(June 1990): 725. For an essay urging historians of technology to decenter or abandon "progress as a conceptual pivot for research," see Philip Scranton, "Determinism and Indeterminacy in the History of Technology," in *Does Technology Drive History? The Dilemma of Technological Determinism*, ed. Merritt Roe Smith and Leo Marx (Cambridge, Mass., 1994), 148.

26. A considerable portion of Edwards's book deals with developments in artificial intelligence and what he calls the "cyborg discourse," which I have not discussed here.

27. Edwards, *Closed World*, ix, 7, 34, 41. For the social constructivist approach, see, for example, Wiebe E. Bijker, *Of Bicycles, Bakelites, and Bulbs: Toward a Theory of Sociotechnical Change* (Cambridge, Mass., 1995). For a sharp critique, see Langdon Winner, "Upon Opening the Black Box and Finding It Empty: Social Constructivism and the Philosophy of Technology," *Science, Technology and Human Values* 18 (Summer 1993): 362–78. Abbate describes social constructionism and systems theory as the key influences on her work. "From ARPANET," 7.

28. For a general discussion of the centrality of military funding to postwar American science and technology, see Stuart W. Leslie, *The Cold War and American Science: The Military-Industrial-Academic Complex at MIT and Stanford* (New York, 1993). See also such works as Everett Mendelsohn, Merritt Roe Smith, and Peter Weingart, eds., *Science, Technology, and the Military* (Dordrecht, The Netherlands, 1988); David Noble, *Forces of Production: A Social History of Automation* (New York, 1984); Merritt Roe Smith, ed., *Military Enterprise and Technological Change: Perspectives on the American Experience* (Cambridge, Mass., 1985); Ann Markusen et al., *The Rise of the Gunbelt: The Military Remapping of Industrial America* (New York, 1991), and the issue of *Osiris* 7 (1992) on "Science after '40," edited by Arnold Thackray.

29. Quoted in Edwards, *Closed World*, 65.

30. Edwards, *Closed World*, 44. He did, however, read the unpublished 1992 report that was the basis of the Norberg and O'Neill book.

31. For another account that persuasively undercuts the inevitability or "obviousness" of the triumph of digital over analog computing, see Larry Owens, "Where are We Going, Phil Morse? Changing Agendas and the Rhetoric of Obviousness in the Transformation of Computing at MIT, 1939–1957," *IEEE Annals of the History of Computing* 18, no. 4 (1996): 34–41. Owens offers a number of nontechnical reasons for the triumph of digital computing, including "Cold War worries about unrest, uncertainty, and unpredictability [that] fed a countervailing emphasis on management and control" (38).

32. Edwards, *Closed World*, 7.

33. Edwards, *Closed World*, 3–4.

34. Hafner and Lyon, *Where*, 29, 34. They dedicate their book to Licklider's memory.

35. J. C. R. Licklider, "Man-Computer Symbiosis," *IRE Transactions on Human Factors in Electronics*, vol. HFE-1 (March 1960): 5; Edwards, *Closed World*, 272; J. C. R. Licklider and Albert Vezza, "Applications of Information Networks," *Proceedings of the*

IEEE 66 (November 1978): 1335. Licklider later told an interviewer that he had "this positive feeling toward the military. It wasn't just to fund our stuff but they really needed it and they were good guys." Edwards, *Closed World*, 267.

36. Edwards, *Closed World*, 101.

37. Norberg and O'Neill, *Transforming*, 270.

38. Although Edwards devotes little attention to counterdiscourses, he does note the "survival" in the current moment of "vestiges" of a "green-world discourse," which he locates in "animistic religions, feminist witchcraft, certain Green political parties, and the deep ecology movement," but he says these "lie at the farthest margins of politics, society, and culture." He argues (and it is an argument that I have trouble following) that "the only possibility for genuine self-determination, is the political subject position of the cyborg." Edwards, *Closed World*, 350.

39. Statement reproduced in Union of Concerned Scientists, 1993 *Annual Report* (Cambridge, Mass., 1994), inside front cover. See also undated flyer, "The Beginnings," from *Union of Concerned Scientists*, Cambridge, Mass., and Leslie, *Cold War and American Science*, 233–41.

40. In the aftermath of demonstrations against military research at the Stanford Research Institute, one group of graduate students, under faculty sponsorship, organized a course on sponsored research at Stanford, which sought to understand "how a generation of close interaction with the Department of Defense has affected Stanford as an academic institution." Quoted in Leslie, *Cold War and American Science*, 248. The group published two volumes on Defense Department research at Stanford. More generally (and from a critical vantage), Brook Hindle argues that "darkside" views of science and technology emerged out of radical protests of the 1960s. Hindle, "Historians of Technology and the Context of History," in Cutcliffe and Post, *In Context*, 235–40.

41. Leslie, *Cold War and American Science*, 245.

42. Steven Levy, *Hackers: Heroes of the Computer Revolution* (1984; reprint, New York: Delta, 1994), 416–18.

43. David Hudson offers a similar "bottom up" perspective on the Net's history in *Rewired*, 13–35.

44. Campbell-Kelly and Aspray, *Computer*, 293.

45. Michael Hauben and Ronda Hauben, *Netizens: On the History and Impact of Usenet and the Internet* (Los Alamitos, Calif., 1997), 41.

46. Hauben and Hauben, *Netizens*, 172; Campbell-Kelly and Aspray, *Computer*, 221. Unix was initially developed at AT&T's Bell Labs in the late 1960s. Although the system was a commercial development, AT&T was prevented by a 1956 consent decree from profiting from sources other than the phone business. As a result, they made Unix widely and cheaply available, and by the 1970s, it became a widely used standard, particularly in academic computing, where a university license cost only $150.

47. Hafner and Lyon, *Where*, 187–218.

48. Hauben and Hauben, *Netizens*, 48–49, x. The second quote comes from a preface signed separately by Michael Hauben. The other chapters appear to have been individually written by Ronda and Michael (who are mother and son), and Michael's chapters tend to take a more aggressively populist stance.

49. Hauben and Hauben, *Netizens*, 102–5.

50. Stephen D. Crocker, "The Origins of RFCs," in J. Reynolds and J. Postal, *RFC 1000: The Request for Comments Reference Guide, August 1987*, http://info.internet.isi.edu:80/in-notes/rfc1000.txt; Hauben and Hauben, *Netizens*, 103, 106–7. The most detailed discussion of the RFCs can be found in Salus's more technically oriented history: *Casting the Net*. Many of the RFCs can be found online at pages maintained by the University of Southern California's Information Sciences Institute: http://www.isi.edu/rfc-editor.org/rfc.html.

51. Levy, *Hackers*, 7, 168, 172. On Nelson, see Gary Wolf, "The Curse of Xanadu," *Wired* 3 (June 1995): 137 ff.

52. Levy, *Hackers*, 272, 156, 143. In a delightful irony that must have been evident to the people behind Community Memory, the computer used was an XDS-940, but it was also known by its original initials, which were very familiar to 1960s activists—SDS. (The change reflected the takeover of Scientific Data Systems by Xerox Corporation.) The online "Community Memory Discussion List on the History of Cyberspace" is named after the Berkeley project. See http://memex.org/community-memory.html.

53. Levy, *Hackers*, 157–68, 181–87, 196–97, 205–6, 214–17, 237–42, 272–77.

54. Campbell-Kelly and Aspray, *Computer*, 220–221.

55. For the origins of the phrase, see "Free Speech Movement: Do Not Fold, Bend, Mutilate or Spindle," anonymous statement from *FSM Newsletter*, reproduced by *Sixties Project* Web site, http://lists.village.virginia.edu/sixties/HTML_docs/Resources/Primary/Manifestos/FSM_fold_bend.html.

56. Leslie, *Cold War and American Science*, 233–34.

57. The Port Huron statement is available online at: http://lists.village.virginia.edu/sixties/HTML_docs/Resources/Primary/Manifestos/SDS_Port_Huron.html. (The most remarkable statement from a subsequent perspective is its warm embrace of nuclear energy.) For Savio's famous statement, see W. J. Rorabaugh, *Berkeley at War: The 1960s* (New York, 1989), 31. The alternative neo-Luddite strain in New Left and Counterculture thought remains potent today. See, for example, Kirkpartrick Sale, *Rebels Against the Future: The Luddites and Their War on the Industrial Revolution: Lessons for the Computer Age* (Reading, Mass., 1995).

58. Severo Ornstein, one of the key BBN engineers, once wore an antiwar button to a briefing on Arpanet with Pentagon officials. Hafner and Lyon, *Where*, 113. Ornstein went on to become the Chair of Computer Professionals for Social Responsibility. See Severo M. Ornstein, "Computers in Battle: A Human Overview," in *Computers in Battle—Will They Work?*, ed. David Bellin and Gary Chapman (Boston, 1987), 1–43.

59. Hauben and Hauben, *Netizens*, 40.

60. These works devote surprisingly little attention to analyzing the obvious role of gendered concepts and practices in a development in which the key figures were almost entirely men. Edwards does offer an interesting analysis of the gendered language of "hard" and "soft" sciences and approaches. Edwards, *Closed World*, 167–73. See also his essay, "The Army and the Microworld: Computers and the Militarized Politics of Gender," *Signs* 16, no. 1 (1990): 102–127.

61. Hafner and Lyon, *Where*, 210; Hauben and Hauben, *Netizens*, 41. See also Campbell-Kelly and Aspray, *Computer*, 292.

62. Hafner and Lyon, *Where*, 240. On NSF and Internet, see David Roessner et al., "The Role of NSF's Support of Engineering in Enabling Technological Innovation," First Year Final Report, January 1997, prepared for the National Science Foundation, http://www.sri.com/policy/stp/techin/.

63. Ian Hardy, "The Evolution of ARPANET Email," unpublished senior thesis, University of California, Berkeley, 1996, http://www.ifla.org/documents/internet/hari1.txt. On the "informalization" of American society in the 1960s, see Kenneth Cmiel, "The Politics of Civility," in *The Sixties: From Memory to History, ed.* David Farber (Chapel Hill, N.C., 1994), 263–90.

64. Thomas Frank, *The Conquest of Cool: Business Culture, Counterculture, and the Rise of Hip Consumerism* (Chicago, 1997), 13.

65. For a detailed discussion of the links between the drug culture and the contemporary computer industry, see Douglas Rushkoff, *Cyberia: Life in the Trenches of Hyperspace* (San Francisco, 1994). According to Rushkoff, programmers regularly circulate lists of which companies are "friendly" to drug users and don't do drug testing (30).

66. This widely repeated phrase was first used (in print) by Stewart Brand in *The Media Lab: Inventing the Future at M.I.T.* (New York, 1987), 202. Less widely used is his corollary that "information also wants to be expensive"—"free" because "it has become so cheap to distribute, copy, and recombine" and "expensive" because "it can be immeasurably valuable to the recipient."

67. Levy, *Hackers*, 229, 268.

68. BBN's entry into commercial networking was spurred by competition from three of their own engineers, who created Packet Communications Incorporated (and demanded the IMP source code). Some companies like Tymshare, which were in the time-sharing business, became network providers; large communications companies like Western Union and MCI also started to offer e-mail. Hafner and Lyon, *Where*, 232–34; Campbell-Kelly and Aspray, *Computer*, 295.

69. IBM charged as much as $300,000 for processors to link its mainframes using its proprietary Systems Network Architecture (SNA). In the 1990s, routers using TCP/IP, which cost a fraction of the price, displaced SNA.

70. On technolibertarianism, see, for example, Paulina Borsook, "Cyberselfish,"

Mother Jones (July/August 1996):56, http://www.motherjones.com/mother_jones/ JA96/borsook.html; Hudson, *Rewired*, 173–259.

71. Mark Lilla, "A Tale of Two Reactions," *New York Review of Books* 45 (May 14, 1998): 7.

72. On WorldCom, see Thomas E. Weber and Rebecca Quick, "Would World-Com-MCI Deal Turn the Net into a Toll-road?" *San Diego Union-Tribune,* October 7, 1997, 11 (originally published in *Wall Street Journal*); Michelle V. Rafter, "WorldCom Bids For No. 1 Status," *Internet World* (October 6, 1997), and Barbara Grady, "Opposition Mounts to WorldCom-MCI Merger," *Internet World* (March 23, 1998), http:// www.iw.com/print/current/index.html. A major subsidiary of WorldCom and the world's largest Internet Service Provider is UUNET, which was founded in 1987 by the academic Unix user's group, Usenix, to sell access to Usenet; it later became a for-profit corporation and was bought by WorldCom in 1996. On the creation of UU-NET, see Salus, *Casting the Net*, 177–78. The counterargument against monopolization of the Internet backbone is the rapid construction of new fiber-optic cables by companies like Qwest. In response to European and American regulatory pressures, MCI sold off its Internet backbone to the British company Cable & Wireless. But some Internet Service Providers still believe that MCI WorldCom will "wield too much power." Arik Hesseldahl, *Internet World* (June 15, 1998). For estimates of control of Internet backbone, see "Top Internet Backbone Companies," *Business Week* (July 20, 1998), www.businessweek.com/1998/29/b3587/123.htm.

73. See Graphic, Visualization, and Usability Center of Georgia Tech, "8th WWW User Survey" (Dec. 1997), reported at http://www.gvu.gatech.edu/user_surveys/ survey-1997–10/#highsum.

74. On Linux, see Glyn Moody, "The Greatest OS That (N)ever Was," *Wired* 5, no. 8 (August 1997): 122 ff and http://www.li.org/. On the Free Software Foundation and Stallman, see its Web pages at http://www.gnu.org/fsf/fsf.html and Richard Stallman, "Why Software Should Not Have Owners," http://www.gnu.org/philosophy/why-free.html. Andrew Leonard, "Apache's Free-Software Warriors!" *Salon* (Nov. 20, 1997), http://archive.salon.com/21st/feature/1997/11/cov_20feature.html. Torvald's Usenet postings are archived at http://x5.dejanews.com/profile.xp?author= torvalds@cs.helsinki.fi%20 (Linus%20Torvalds).

75. Phil Agre, "The Internet and Public Discourse," *First Monday* 3 (March 2, 1998), www.first-monday.dk/issues/issue3_3/agre/index.html.

11. The Road to Xanadu: Public and Private Pathways
on the History Web

1. T. H. Nelson, "A File Structure for the Complex, the Changing, and the Indeterminate," *Proceedings of the 20th ACM National Conference* (1965): 84–100. Nelson's

ideas about hypertext were heavily influenced by Vannevar Bush, "As We May Think" (1945); for a reprint of the article and discussions of its influence, see James M. Nyce and Paul Kahn, ed., *From Memex to Hypertext: Vannevar Bush and the Mind's Machine* (Boston, 1991). Even earlier, in 1938, H. G. Wells talked of creating a "World Encyclopedia" with a true "planetary memory for all mankind": quoted in Michael Lesk, "How Much Information Is There in the World?," unpublished paper, 1997, http://www.lesk.com/mlesk/ksg97/ksg.html. (Unless otherwise noted, the Web references in this article were rechecked online on May 5, 2001.)

2. Theodor Holm Nelson, "Xanalogical Structure, Needed Now More than Ever: Parallel Documents, Deep Links to Content, Deep Versioning, and Deep Re-Use," *ACM Computing Surveys* 31 (Dec. 1999), http://www.cs.brown.edu/memex/ACM_HypertextTestbed/papers/60.html; see also Ted Nelson, "Who I Am: *Designer, Generalist, Contrarian* Theodor Holm Nelson, 1937–," http://www.sfc.keio.ac.jp/~ted/TN/WhoIAm.html; and Theodor Holm Nelson, "Opening Hypertext: A Memoir," in *Literacy Online: The Promise (and Peril) of Reading and Writing with Computers,* ed. Myron C. Tuman (Pittsburgh, 1992), 43–57.

3. Gary Wolf, "The Curse of Xanadu," *Wired* 3 (June 1995), http://www.wirednews.com/wired/archive/3.06/xanadu_pr.html; Theodor Holm Nelson, "Errors in 'The Curse of Xanadu,' by Gary Wolf," in Andrew Pam, *Xanadu Australia*, http://www.xanadu.com.au/ararat.

4. For a history of the development of the Internet, see John Naughton, *A Brief History of the Future: From Radio Days to Internet Years in a Lifetime* (Woodstock, 2000), 229–63.

5. For detailed information on Web search engines, see the materials at Search Engine Watch, http://www.searchenginewatch.com/. Search Engine Watch and other commentators currently rate Google the best overall Web search tool.

6. U.S. Department of Commerce, *The Emerging Digital Economy* (Washington, 1998), quoted in Stephen Segaller, *Nerds 2.01: A Brief History of the Internet* (New York, 1998), 14. "Sizing Up the Web," *New York Times*, Dec. 11, 2000, C4. All *New York Times* articles cited here are available online (generally for a per-article fee of $2.50) at "The *New York Times* on the Web," http://www.nytimes.com and (for a library subscription fee) through Lexis-Nexis Academic Universe, http://web.lexis-nexis.com/universe; where a page number is cited, the article was first consulted in the print version of the *Times*; where a specific URL (uniform resource locator) is cited, the article is available online for free. Office of Research, OCLC (Online Computer Library Center, Inc.), "Web Statistics," in Web Characterization Project, http://wcp.oclc.org/. Google, http://www.google.com. Peter Lyman and Hal R. Varian, "How Much Information?," *Journal of Electronic Publishing* 6 (Dec. 2000), http://www.press.umich.edu/jep/06-02/lyman.html. BrightPlanet, "The Deep Web: Surfacing Hidden Value," in BrightPlanet.com, *Complete Planet*, http://www.completeplanet.com/Tutorials/

DeepWeb/index.asp; Lisa Guernsey, "Mining the 'Deep Web' with Specialized Drills," *New York Times*, Jan. 25, 2001.

7. The Internet Archive, http://www.archive.org, intends "to permanently preserve a record of public material" on the Internet. At the present time, however, use of their archive requires programming skills, and I did not receive a response to the request to use the archive that I submitted in October 2000. For a discussion of the need to archive the Web (and a complaint about lack of response from the Internet Archive), see Richard Wiggins, "The Unnoticed Presidential Transition: Whither Whitehouse. gov?," *First Monday* 6 (Jan. 8, 2001), http://www.firstmonday.org/issues/issue6_1/ Wiggins/index.html. Michael O'Malley and Roy Rosenzweig, "Brave New World or Blind Alley? American History on the World Wide Web," this volume.

8. See "Collections Currently in Progress," in Library of Congress, American Memory: Historical Collections for the National Digital Library, http://memory.loc. gov/ammem/amfuture.html. See, more generally, Committee on an Information Technology Strategy for the Library of Congress of National Research Council, *LC21: A Digital Strategy for the Library of Congress* (Washington, 2000) http://books. nap.edu/books/0309071445/html/index.html. As of December 2000, the NDLP had 5,772,967 items online, but some American Memory materials are available as a result of the Ameritech Program and others as a result of cooperative agreements with other institutions. NDLP Reference Team to Roy Rosenzweig, e-mails, Feb. 15, 2001.

9. Peter R. Henriques, "The Final Struggle between George Washington and the Grim King: Washington's Attitude toward Death and an Afterlife," *Virginia Magazine of History and Biography* 107 (Winter 1999): 75, 95–96. Henriques discussed his methodology with Rosenzweig on November 6, 2000.

10. OCLC, "Web Statistics"; Peter B. Hirtle, "Free and Fee: Future Information Discovery and Access," *D-Lib Magazine* 7 (Jan. 2001), http://www.dlib.org/dlib/january01/01editorial.htm.

11. Kevin M. Guthrie, "Revitalizing Older Published Literature: Preliminary Lessons from the Use of JSTOR," paper presented at the conference "Economics and Usage of Digital Library Collections," Ann Arbor, March 23–24, 2000, http://www. jstor.org/about/preliminarylessons.html. See also "Editor's Interview: Developing a Digital Preservation Strategy for JSTOR, an interview with Kevin Guthrie," *RLG DigiNews* 4, no. 4 (2000), http://www.rlg.org/preserv/diginews/deginews4-4.html# feature1. John Spargo, "The Influence of Karl Marx on Contemporary Socialism," *American Journal of Sociology* 16 (July 1910): 21–40. Fred Shapiro's discoveries are discussed in Ethan Bronner, "You Can Look It Up, Hopefully," *New York Times*, Jan. 10, 1999, http://www.nytimes.com/library/review/011099language-database-review.html.

12. Barbara Quint, "Gale Group's InfoTrac OneFile Creates Web-Based Periodical Collection for Libraries," *Information Today NewsBreaks*, Oct. 16, 2000, http://www. infotoday.com/newsbreaks/breaks.htm.

13. See http://dir.yahoo.com/Arts/Humanities/History/; "Search Websites" in Center for History and New Media, http://chnm.gmu.edu/chnm/websites.taf.

14. *Choice* quoted in William G. Thomas and Alice E. Carter, *The Civil War on the Web: A Guide to the Very Best Sites* (Wilmington, 2000), xiii; Library of Congress, American Memory, http://memory.loc.gov/; National Park Service, Links to the Past, http://www.cr.nps.gov; Virginia Center for Digital History, http://www.vcdh .virginia.edu/; Thomas R. Fasulo, Battle of Olustee, http://extlab1.entnem.ufl.edu/ olustee/; Scott McKay, Official Historic Website of the 10th Texas Infantry, http:// members.aol.com/SMckay1234/.

15. "Facts and Statistics," in Family Search, Church of Jesus Christ of Latter-day Saints http://www.familysearch.com/Eng/Home/News/frameser_news.asp?PAGE= home_facts.asp.

16. Ibid. April Leigh Helm and Matthew L. Helm, *Genealogy Online for Dummies* (New York, 1999).

17. Diane Ravitch, ed., *The American Reader: Words That Moved a Nation* (New York, 1990); Elizabeth Cady Stanton, "The Solitude of Self," in *American Public Address, 1644–1935,* University of Arkansas Supplement to Communication 4353, Bernadette Mink, http://comp.uark.edu/~brmink/stanton.html; "Niagara Movement Declaration of Principles, 1905" in American History Class Enhancement Pages, Thomas Martin, http://www.sinclair.edu/classenhancements/his101e-tm/civilrt1.htm; M. Carey Thomas, "Higher Education for Women," in Mrs. Pojer's History Classes' Home Page, Susan M. Pojer, http://zuska.simplenet.com/USProjects/DBQs2000/ APUSH-DBQ-40.htm. The last two sites were accessed in October 2000 but were no longer available in May 2001. In the first instance, the material moved to a gated WebCT server.

18. W. E. Burghart Du Bois, "The Talented Tenth," in Mr. Kenyada's Neighborhood, Richard Kenyada, http://www.kenyada.com/talented.htm. Joe Hill, "The Preacher and the Slave," in *History in Song,* Manfred J. Helfert, http://www.fortunecity.com/tin-pan/parton/2/pie.html. Dean B. McIntyre, "'Lift Every Voice'—100 Years Old," in *General Board of Discipleship,* United Methodist Church, http://www.gbod.org/worship/ default.asp?act=reader&item_id=1786. Alice Duer Miller, "Evolution," in *poet ch'i,* Kevin Taylor, http://www.geocities.com/Paris/Bistro/8066/index2.htm. Carl Becker, "Everyman His Own Historian," *American Historical Review* 37 (Jan. 1932): 221–36.

19. BoondocksNet.com, http://www.BoondocksNet.com; Jim Zwick to Rosenzweig, e-mails, Nov. 1, 27, 2000. Some scholars will face copyright and archival restrictions in placing their research materials online, but a surprisingly large percentage of materials that historians use—books, magazines, and newspapers from before 1923 and government documents, for example—are in the public domain.

20. "What Is H-Net?," in *H-Net: Humanities & Social Sciences OnLine,* MATRIX: The Center for Humane Arts, Letters, and Social Sciences OnLine, Michigan State University, http://www2.h-net.msu.edu/about/.

21. Thomas and Carter, *Civil War on the Web*, xvi–xix; Golden Ink, *About North Georgia*, http://ngeorgia.com, quoted ibid., xix.

22. Elizabeth Cady Stanton, *Solitude of self: address delivered by Mrs. Stanton before the Committee of the Judiciary of the United States Congress, Monday, January 18, 1892* (Washington, 1915), in Rare and Special Collections Division, Library of Congress, *Votes for Women: Selections from the National American Woman Suffrage Association Collection, 1848–1921*, http://lcweb2.loc.gov/ammem/naw/nawshom.html.

23. Voice of America, *The Century in Sound: An American's Perspective*, http://www.voa.gov/century/century.html; "Socialist Eugene V. Debs Speaks During the Presidential Campaign of 1904," in *Eyewitness: History through the Eyes of Those Who Lived It*, Ibis Communications, Inc., http://www.ibiscom.com/vodebs.htm; "Eugene V. Debs," in *Pluralism and Unity*, David Bailey, David Halsted, and Michigan State University, http://www.expo98.msu.edu/sounds/debs.html. The voice is correctly identified as that of an actor in the Department of History, University at Albany, State University of New York, U.S. Labor and Industrial History World Wide Web Audio Archive, http://www.albany.edu/history/LaborAudio/. For a discussion of the provenance of the Debs speech, see Roy Rosenzweig and Stephen Brier, *Who Built America? From the Centennial Celebration of 1876 to the Great War of 1914* (CD-ROM) (New York, 1993), 352.

24. See, for example, "The Willie Lynch Speech of 1712," in *Shepp's Place*, Will Shepperson, http://www.eden.rutgers.edu/~wshepp3/lynch.html; and Willie Lynch, "How to Control the Black Man for at least 300 Years," in KohlBlackTimes.com, http://www.kohlblacktimes.com/willie.htm. The best online commentary on the Lynch speech is Anne Cleëster Taylor, "The Slave Consultant's Narrative: The Life of an Urban Myth?," in *African Missouri*, Anne Cleëster Taylor, http://www.umsl.edu/~libweb/blackstudies/narrate.htm. See also Mike Adams, "In Search of Willie Lynch," *Baltimore Sun*, Feb. 22, 1998, 1 (available online in Lexis-Nexis Academic Universe). Of course, many real documents make points similar to those in the Lynch speech.

25. For a discussion of the inclusiveness of virtual libraries, see James J. O'Donnell, *Avatars of the Word: From Papyrus to Cyberspace* (Cambridge, Mass., 1998), 29–43.

26. Kendra Mayfield, "Library of Congress Goes Digital," *Wired News*, Jan.19, 2001, http://www.wired.com/news/print/0,1294,41166,00.html. For list of sponsors, see "A Unique Public-Private Partnership Supporting the National Digital Library," in American Memory, Library of Congress, http://memory.loc.gov/ammem/sponsors.html. See "Library of Congress/Ameritech National Digital Library Competition," ibid., http://memory.loc.gov/ammem/award/index.html.

27. For an astute discussion of Valley of the Shadow, http://jefferson.village.virginia.edu/vshadow/, see Gary J. Kornblith, "Venturing into the Civil War, Virtually: A Review," *Journal of American History* 88 (June 2001): 145–51, http://www.historycooperative.org/journals/jah/88.1/kornblith.html.

28. Franklin and Eleanor Roosevelt Institute and Institute for Learning Technologies, New Deal Network, http://newdeal.feri.org/; Center for History and New Media and American Social History Project, *History Matters: The U.S. Survey Course on the Web*, http://historymatters.gmu.edu. *History Matters* also includes annotated lists of history Web sites, online assignments, interactive exercises on the historian's craft, and teaching forums with leading scholars and teachers.

29. University of North Carolina Libraries, *Documenting the American South*, http://docsouth.unc.edu/aboutdas.html.

30. Special Projects Program in the Information and Intelligent Systems Division of the Directorate for Computer and Information Science Engineering, National Science Foundation, Digital Libraries Initiative, http://www.dli2.nsf.gov/. Other agencies such as NEH participated in the initiative, but most of the money came from NSF. Mark Lawrence Kornbluth et al., National Gallery of the Spoken Word, http://www.ngsw.org/.

31. Wendy Lougee to Rosenzweig, e-mail, Nov. 3, 2000; Maria Bonn, project director for MOA, provided helpful information on the project in a phone conversation with Rosenzweig, Nov. 9, 2000.

32. Steven Gelber quoted in Nancy Ross-Flanigan, "The Making of America," *Michigan Today* (Spring 1998), http://www.umich.edu/~newsinfo/MT/98/Spr98/mt15s98.html.

33. "Thoughtful weeding of reformatted material is a necessary element of an overall collection management program in the nation's major research libraries": University of Michigan Digital Library Production Service, "Principles and Considerations for University of Michigan Library Subject Specialists" (Feb. 2000), http://www.umdl.umich.edu/policies/digitpolicyfinal.html. Nicholson Baker, "Deadline: The Author's Desperate Bid to Save America's Past," *New Yorker*, July 24, 2000, 42–61. See also Nicholson Baker, *Double Fold: Libraries and the Assault on Paper* (New York, 2001).

34. Association of Research Libraries, "Talking Points in Response to Nicholson Baker's Article in the 24 July *New Yorker*," http://www.arl.org/scomm/baker.html. See also Barbara Quint, "Don't Burn Books! Burn Librarians!! A Review of Nicholson Baker's *Double Fold: Libraries and the Assault on Paper*," *Searcher* 9, no. 6 (June 2001), http://www.infotoday.com/searcher/jun01/voice.htm. Thanks to Josh Brown for his help with this issue. Searching by the word is only possible where the text has been converted into codes that the computer understands as letters and words. The term "digitizing" can refer confusingly both to scanning an image of a page of text and to converting those images of letters into codes that the computer can understand as letters. It is relatively easy to scan thousands of pages of text as images; it is much harder to get that into machine-readable form. That requires either retyping or an OCR (optical character recognition) system. MOA uses an automated OCR system, which gives very good but not perfect results.

35. Gelber quoted in Ross-Flanigan, "Making of America." Association of Research Libraries, "Summary of Fiscal Year 1999 Appropriation Request for the National Endowment for the Humanities," in *Association of Research Libraries*, http://www.arl.org/info/letters/FY1999.html; Stanley N. Katz, "Rethinking the Humanities Endowment," *Chronicle of Higher Education*, Jan. 5, 2001, B5–10. All *Chronicle* articles cited here are available online to subscribers at http://chronicle.com/weekly/sitesearch.htm; where a page number is cited, the article was first consulted in the print version of the *Chronicle*.

36. *LC21*; James O'Donnell quoted in Katie Hafner, "Saving the Nation's Digital Legacy," *New York Times*, July 27, 2000, G1. See also Mayfield, "Library of Congress Goes Digital."

37. Daren Fonda, "Copyright's Crusader," *Boston Globe Magazine*, Aug. 29, 1999, quoted in Dennis S. Karjala, *Opposing Copyright Extension*, http://www.public.asu.edu/~dkarjala/commentary/Fonda8–29–99.html. See, for example, *NCC Washington Update*, March 27, 1998, http://www2.h-net.msu.edu/~ncc/ncc98/ncc9811mar27.html. Roy Rosenzweig and Stephen Brier, *Who Built America? From the Centennial of 1876 to the Great War of 1914* (CD-ROM); Roy Rosenzweig et al., *Who Built America? From the Great War of 1914 to the Dawn of the Atomic Age in 1946* (CD-ROM) (New York, 2000).

38. Kathy Perry, director of VIVA, provided information to Rosenzweig in several conversations during December 2000 and January 2001.

39. Contentville, http://www.contentville.com/.

40. Corbis and Getty "have been gobbling up smaller agencies around the world": Gordon Black, "Corbis Courts Online Consumers," *Seattle Times*, Nov. 16, 1999, D6. See also Kristi Heim, "Digital Image is Everything as Gates, Getty Vie for Control of 'Net Art," *Denver Post*, March 5, 2000, I-03 (both available online through Lexis-Nexis Academic Universe). Corbis Corporation, Corbis—The Place for Pictures Online, http://www.corbis.com.

41. EBSCO's full-text holdings in history do not appear to be as deep as those from ProQuest and EAA. For example, EBSCO does not offer such standards as *Journal of Women's History*, *Journal of Negro History*, and *Journal of Southern History*, which are in EAA.

42. On the electronic book ventures, see Goldie Blumenstyk, "Digital-Library Company Plans to Charge Students a Monthly Fee for Access," *Chronicle of Higher Education*, Nov. 14, 2000; Andrew R. Albanese, "E-Book Gold Rush: Welcome to the Electronic Backlist," *Lingua Franca* 10 (Sept. 2000), http://www.linguafranca.com/print/0009/inside-ebook.html; Jennifer Darwin, "Storybook Beginning: Questia-Founder Follows Novel Script to Launch Online College Library," *Houston Business Journal*, April 7, 2000, http://www.bizjournals.com/houston/stories/2000/04/10/story2.html; Lisa Guernsey, "The Library as the Latest Web Venture," *New York Times*, June 15, 2000; LC21, box 1.3; Tom Fowler, "$90 Million in Funding for Questia,"

Houston Chronicle, Aug. 24, 2000, business p. 1 (available online in Lexis-Nexis Academic Universe); and Kendra Mayfield, "The Quest for E-Knowledge," *Wired News*, Feb. 5, 2001, http://www.wired.com/news/print/0,1294,41543,00.html. For survey, see David Thelen, "The Practice of American History," *Journal of American History* 81 (Dec. 1994): 953. History is not particularly well represented in the NetLibrary collection so far. Some other "e-book" vendors concentrate on particular fields, for example, information technology (ITKnowledge) and marketing and finance (Books24x7).

43. See HarpWeek, "Purchase Information," in HarpWeek, http://www.harpweek.com/4AboutHarpWeek/HowToPurchase/HowToPurchase.htm. HarpWeek may also begin levying annual maintenance fees in 2002.

44. Robert Thibadeau to Rosenzweig, e-mails, Nov. 1, 2, 2000; The Historical New York Times Project, http://nyt.ulib.org/. For unreadable pages, see, for example, Aug. 6, 1860, and Aug. 6, 1863.

45. Michael Jensen, "Mission Possible: Giving It Away While Making It Pay," paper presented at the annual meeting of the Association of American University Presses, Austin, Tex., June 22, 1999, http://www.nap.edu/staff/mjensen/aaup99.html (emphasis in original).

46. On the History Cooperative, see Michael Grossberg, "Devising an Online Future for Journals of History," *Chronicle of Higher Education*, April 21, 2000. *William and Mary Quarterly*, *Western Historical Quarterly*, *History Teacher*, and *Law and History Review* will soon join the *Journal of American History* and the *American Historical Review* in the History Cooperative. (Full disclosure: I was a member of the *Journal of American History* committee that developed the cooperative project.)

47. For an experiment in hypertext publishing, see the articles in Roy Rosenzweig, ed., "Hypertext Scholarship in American Studies," http://chnm.gmu.edu/aq; and Roy Rosenzweig, ed., "Forum on Hypertext Scholarship: *AQ* as Web-Zine: Responses to *AQ*'s Experimental Online Issue," *American Quarterly* 51 (June 1999): 237–82 (available online to subscribers at Project Muse, http://muse.jhu.edu/).

48. "arXiv Monthly Submission Rate Statistics," in arXiv.org e-Print archive, http://arXiv.org/cgi-bin/show_monthly_submissions; Stevan Harnad, "The Future of Scholarly Skywriting," in "*i* in the Sky: Visions of the Information Future," ed. A. Scammell, Aslib, Nov. 1999, http://www.cogsci.soton.ac.uk/~harnad/Papers/Harnad/harnad99.aslib.html. See also Vincent Kiernan, "Open Archives' Project Promises Alternative to Costly Journals," *Chronicle of Higher Education*, Dec. 3, 1999; Herbert Van de Sompel and Carl Lagoze, "The Santa Fe Convention of the Open Archives Initiative," *D-Lib Magazine* 6 (Feb. 2000), http://www.dlib.org/dlib/february00/vandesompel-oai/02vandesompel-oai.html; Steven Harnad, "Free at Last: The Future of Peer-Reviewed Journals," *D-Lib Magazine* 5 (Dec. 1999), http://www.dlib.org/dlib/december99/12harnad.html.

49. Jensen, "Mission Possible."

50. David D. Kirkpatrick, "Librarians Unite against Cost of Journals," *New York Times*, Dec. 25, 2000, C5. Data on library budgets provided by Mary Case of the As-

sociation of Research Libraries and published in *ARL Statistics, 1998–99* (Washington, 2000); *ARL Supplementary Statistics, 1998–99* (Washington, 2000). On the crisis in scholarly publishing, see, for example, Sanford G. Thatcher, "Thinking Systematically about the Crisis in Scholarly Communication," and other papers presented at the conference "The Specialized Scholarly Monograph in Crisis; or, How Can I Get Tenure If You Won't Publish My Book?," Washington, Sept. 11–12, 1997, http://www.arl.org/scomm/epub/papers/; and Roy Rosenzweig, "How Can I Get Tenure If You Won't Publish My Book?," *Organization of American Historians Newsletter* 29 (Nov. 1997): 5.

51. Christopher Stern, "Freelancers Get Day in Court," *Washington Post*, Nov. 7, 2000, E3. David D. Kirkpatrick, "Publisher Set to Split E-Book Revenue," *New York Times*, Nov. 7, 2000, C2.

52. The National Coalition of Independent Scholars (NCIS) successfully lobbied the Modern Language Association to pass two resolutions on access for independent scholars at their December 2000 annual meeting in Washington, D.C. See Margaret Delacy, "A History of NCIS," http://www.ncis.org/history.htm; and NCIS, "Modern Language Association Overwhelmingly Passes Resolutions Concerning Access to Resources by Independent Scholars," http://www.ncis.org/ncisnews.htm#MLA%20 Convention, both in *National Coalition of Independent Scholars*, http://www.ncis.org.

53. On network effects and economies of scale, see Philip E. Agre, "The Market Logic of Information," paper presented at Interface 5, Sept. 2000; Carl Shapiro and Hal Varian, *Information Rules: A Strategic Guide to the Network Economy* (Boston, 1998); and Philip E. Agre, "Notes and Recommendations," *Red Rock Eater Digest*, March 3, 1998, http://commons.somewhere.com/rre/1998/notes.and.recommendation2.html.

54. "State Has Eight Firms on Forbes' List of Biggest 500 Private," *Associated Press State & Local Wire*, Nov. 16, 2000 (available in Lexis-Nexis Academic Universe); "EBSCO Publishing Corporate Quick Facts," in EBSCO Publishing Homepage, http://www.epnet.com/bground2.html. UMI is considering plans to turn the page images into searchable text, potentially a massive project. Paula J. Hane, "UMI Announces Digital Vault Initiative," *Information Today, Newsbreak*, July 13, 1998, http://www.infotoday.com/newsbreaks/nb0713-3.htm. For a report that digital facsimiles will be provided, see "*Times* Pages to Be Available on Internet," *New York Times*, Jan. 13, 2001, http://www.nytimes.com/2001/01/13/technology/13BELL.htm. I have heard reports that the pages will ultimately be converted to searchable form through a combination of OCR and retyping of headlines and first paragraphs.

55. BigChalk: The Education Network, http://www.bigchalk.com.

56. On the Internet boom, see Hal R. Varian, "Economic Scene," *New York Times*, Feb. 6, 2001, C2. Lightspan.com, http://www.lightspan.com/. As of January 2001, most of the links to materials in history said: "We're currently gathering the best educational links for this topic. Soon, you'll have access to expert-selected Web sites, encyclopedia articles, learning activities, lesson plans, and more."

57. The list of best Web sites was not officially launched when I viewed it on February 6, 2001, but it already contained a long list of Civil War sites. "The History Channel.Com Network," in The History Channel.com, http://network.historychannel.com/index.asp?page=home. Cowles History Group, Inc., "The HistoryNet: Advertiser Information," in The HistoryNet, http://www.thehistorynet.com/forms/adinfo.htm; "About Us: Our Story," in About.com, About—The Human Internet, http://ourstory.about.com/index.htm?PM=59_1100_T.

58. Oakleigh Thorne quoted in Sarah Carr and Goldie Blumenstyk, "The Bubble Bursts for Education Dot-Coms," *Chronicle of Higher Education*, June 30, 2000, A39–40. "Discovery.Com Workers Get Pink Slips," *Washington Post*, Nov. 14, 2000, C7. "Online Advertising Rate Card Prices and Ad Dimensions," Aug. 14, 2000, in *AdRelevance*, Jupiter Media Metrix, http://www.adrelevance.com/intelligence/intel_archive.jsp; Paul F. Nunes, "Wake-up Call for Internet Firms Overly Dependent on Ad Revenues," *BusinessWorld* (Philippines), June 6, 2000 (available in Lexis-Nexis Academic Universe).

59. See, for example, Roy Rosenzweig, "Marketing the Past: American Heritage and Popular History in the United States," in *Presenting the Past: Essays on History and the Public*, ed. Susan Porter Benson, Stephen Brier, and Roy Rosenzweig (Philadelphia, 1986), 21–49.

60. Susan Smulyan, *Selling Radio: The Commercialization of American Broadcasting, 1920–1934* (Washington, 1994); Stuart Elliott, "Banners' Ineffectiveness Stalls an Up-and-Coming Rival to TV," *New York Times*, Dec. 11, 2000, C4; "Dot Coms in the Driver's Seat," Sept. 5, 2000, in *AdRelevance* http://www.adrelevance.com/intelligence/special_dotcom.pdf; "The Failure of New Media," *Economist*, Aug. 19, 2000.

61. David D. Kirkpatrick, "Media Giants in Joint Deal for Harcourt," *New York Times*, Oct. 28, 2000, C1. See also Richard Poynder, "The Debate Heats Up—Are Reed Elsevier and Thomson Corp. Monopolists?," *Information Today Newsbreaks* (April 30, 2001), http://www.infotoday.com/newsbreaks/nb010430–1.htm.

62. Brett D. Fromson, "On the Level: Is This a Stock 'Primed' for an Uptick?," The Street.com, Dec. 5, 2000, http://www.thestreet.com/_yahoo/markets/onthelevel/1199748.html. (The merger was completed March 1, 2001.) "Primedia's Loss Exceeds Expectations, Taking Hit from New-Media Businesses," *WSJ.Com*, Feb. 2, 2001, http://public.wsj.com/sn/y/SB981035131440666351.html (accessed Feb. 17, 2001, but not accessible on May 5, 2001).

63. Victoria Murphy, "Unlocking the Vault," *Forbes Magazine*, Nov. 13, 2000, http://www.forbes.com/forbes/2000/1113/6613228a.html (*Forbes* now requires that you register to access its articles); Steve Silberman, "Putting History Online," *Wired News*, June 26, 1998, http://www.wired.com/news/culture/0,1284,13298,00.html. See also Péter Jacsó, "With Experience and Content, UMI Is Poised for a Conversion Megaproject," *Information Today*, Sept. 8, 1998, http://www.infotoday.com/it/sep98/jacso.htm and the enhanced version, http://www.umi.com/hp/News/Reviews/Site

Builder.html; "Bell & Howell's ProQuest Digital Vault Initiative Leaps Forward This Spring," press release, March 22, 2000, in Bell & Howell's Learning and Information, http://www.umi.com/hp/PressRel/20000322.html.

64. Neal Stephenson, *Snow Crash* (New York, 1992), 22.

65. Florence Olsen, "'Open Access' is the Wave of the Information Future, Scholar Says," *Chronicle of Higher Education*, Aug. 18, 2000; William Y. Arms, "Automated Digital Libraries: How Effectively Can Computers Be Used for the Skilled Tasks of Professional Librarianship?," *D-Lib Magazine* 6 (July–August 2000), http://www.dlib.org/dlib/july00/arms/07arms.html.

66. For a recent effort by librarians and scientists to fight back against the rapacious prices of commercially owned science journals, see Scholarly Publishing & Academic Resources Coalition and Triangle Research Libraries Network, *Declaring Independence: A Guide to Creating Community-Controlled Science Journals* (Washington, 2001), http://www.arl.org/sparc/DI/.

67. Nelson, "A File Structure for the Complex."

Page numbers in *italics* refer to illustrations.